BEYOND INTELLIGENT DESIGN

From an autonomous universe
to a functional virtual reality

Efthimios Harokopos

Beyond Intelligent Design
From an autonomous universe to a functional virtual reality

Efthimios Harokopos

Book website:
http://www.digitalcosmology.com/Blog/beyond-intelligent-design/

Cover image: NASA/JPL-Caltech, 2006
Target Name: Mira; Mission: Galaxy Evolution Explorer (GALEX)
http://photojournal.jpl.nasa.gov/catalog/PIA09958

ISBN-13: 978-1511400626
ISBN-10: 1511400625

1st edition 2015
Printed in the United States of America

Design, Layout, and Typesetting by Alexander Becker
www.alexanderbecker.net

Table of Contents

Dedicated to the memory of my sister Angeliki.

INTRODUCTION

During the last two decades, a new doctrine has emerged known as intelligent design. This new doctrine has caused a heated debate between its proponents and the proponents of methodological naturalism. Methodological naturalism is the doctrine that science and specifically knowledge acquisition about physical reality should rely only on the scientific method and natural causes without making any hypotheses about the existence of supernatural causes. Intelligent design is a hypothesis that certain features of the world, and specifically of living beings, are best explained by an intelligent cause.[1] According to proponents of methodological naturalism, intelligent design is a hypothesis that has no place in science. Many scientists, including primarily those who accept modern evolution theory, consider intelligent design to be covert creationism. At the same time, proponents of intelligent design claim that they do not deny the value of modern evolution theory and they also argue that their hypothesis has no relation to biblical creation. Let us be bold early on at this point; in essence, this is the same old debate between creationists and evolutionists and only the arguments of the two sides are now different. The majority of scientists accept modern evolution theory and natural selection. Many believe that life started by chance, whereas intelligent design proponents believe that life or at least some important features of living organisms were created by some higher intelligence. This is basically the same old debate between people who believe in God as the creator of human beings and atheists or, in the best case, agnostics.

Contemporary proponents of intelligent design include Philip E. Johnson, Michael Behe, Stephen C. Meyer and William Dembski. Well-known critics of intelligent design include

---

1   Discovery Institute: http://www.discovery.org/

John Wilkins, Richard Dawkins, Wesley Elsberry, Victor Stenger, Elliot Sober, Eugene Scott and Glenn Branch. Most of the critics argue that intelligent design is an argument from ignorance and not a credible scientific thesis.

Let us also not hide from the truth regardless of our personal beliefs regarding intelligent design. Most contemporary scientists are realists and adhere to the philosophical position that all phenomena, including life, can be explained within the framework of physical laws and look towards the development of a *Theory of Everything*. They accuse proponents of intelligent design of trying to introduce supernatural causes in science. In addition, for many contemporary scientists and philosophers, mainly empiricists, an explanation of the causes of phenomena should not be part of scientific inquiry and any reference to causes or laws of nature belongs to metaphysics. In the context of these philosophical positions, intelligent design introduces a cause and hence it is metaphysics and has no place in science.

However, although the main tool of scientists for achieving progress is the scientific method, it turns out that many predictions they put to test arise from theories that make specific metaphysical commitments (Esfeld, 2009). By confirming predictions in laboratories, scientists may be indirectly affirming commitments, known or even unknown to them, to specific metaphysics. It is impossible to even imagine a modern scientific theory that does not make some sort of metaphysical claims about the nature of physical reality. The fact of the matter is that there is no scientific theory that relies on only a few empirical principles that when put together they generate new predictions that are not deduced directly from the phenomena. Modern scientific theories are as hypothetical as their metaphysical claims (ibid.). Although scientific theories in general need not necessarily commit to metaphysical positions, the theories that comprise mainstream science nowadays make such commitments, which often come to conflict with intuition and common sense, as we shall see later in this book.

The unreasonable stand of many supporters of methodological naturalism can be probably explained when one realizes that most scientists are not aware of the metaphysical commitments of their theories. In many cases the metaphysics are concealed in the foundations of the theory and scientists deal only with a set of equations and the predictions that arise from their solutions. One question is then: what is the difference between explicit metaphysics, as in the case of the intelligent design hypothesis and implicit metaphysics, such as in the case of modern scientific theories? One answer is that implicit metaphysics can be obscured, never discussed or even undermined as unimportant, whereas explicit metaphysics can be easily attacked. Thus, ignorance, or even obscurantism, often plays a major role when scientists attack metaphysics. Modern theories generate predictions about entities and features of reality that are either difficult to observe or even unobservable. For example, the Higgs boson and associate field are key elements of a hypothesis about a mechanism of matter formation. However, the difference between such a hypothesis and the hypothesis of intelligent design is that the former fits within the framework of an accepted falsifiable model, known as the standard model of particle physics, whereas the latter does not. Let us keep this difference in mind, because it is important.

The failed case for intelligent design

Intelligent design followers want to make their bold hypothesis part of the educational curriculum without upsetting the scientific establishment. A consequence of such bold efforts was a court ruling in the United States in the Kitzmiller et. al. V Dover Area Scholl District case in 2005.[2] The honor-

---

2   http://en.wikipedia.org/wiki/Kitzmiller_v._Dover_Area_School_District

able judge John E. Jones III ruled that intelligent design is not science but creationism in disguise. I strongly believe that the decision might have been somewhat different had the defense made a case for intelligent design in a proper way. But it appears they did not know how to. Not only they did not present a sound case, but there are reports that the defendants tried to distance themselves from their own position as the court decision stated. This is quite peculiar but also understandable. It is hard to present a case in favor of intelligent design unless one understands the weaknesses in the foundations of science and that modern theories are as hypothetical as their metaphysical commitments.

Intelligent design backers have failed so far to present a credible case in favor of their hypothesis. Arguments based on the existence of irreducibly complex biological systems (Behe, 1996) and complex specified information (Dembski, 2002) have not convinced the scientific community to even consider publishing papers on the subject in peer reviewed journals. In my opinion, the failure is because they have not presented an alternative theory that melds the intelligent design hypothesis, or a variation of it, in the same way most modern theories combine their own metaphysical hypotheses. Intelligent design is a hypothesis that is not falsifiable and this is in addition to the fact that it cannot be proved. There is a way to correct the mistake, but it involves a lot more than just presenting a bold hypothesis. There will be more about this later in this section after the difference between a hypothesis and a scientific theory has been clarified.

Some have claimed that intelligent design is a conclusion of *inductive reasoning,* a process of thought that was at center stage in the early development of *natural philosophy* by Galileo, Newton and their successors. However, this assertion is false because a proposition like, for example, "the eye was designed", is an analytic proposition and not a general rule arrived at from other true proposition or facts. These types of propositions often arise from *abductive reasoning* and they are called *hypotheses*. The only way of using non-falsifiable hypotheses in a deductive theory is to raise them to the status of a postulate, or axiom, and assume that they are true. This is different from what proponents of intelligent design have done so far. They have failed to present a theory in which their hypothesis has the status of a postulate and at the same time this theory generates new predictions that can be tested in laboratories. The fact is that a postulate similar to "the eye was designed", or even one that states that specific features of living organisms were designed, does not add value to any theory but instead it adds redundancy and non-falsifiability.

On the other side of the debate, it is intelligent design opponents that are faced with a serious problem. What they think is the main advantage of evolution theory, i.e., the fact that it is falsifiable, turns out to be also its main disadvantage. This is because there are only two alternatives possible, i.e., either man is the outcome of random particle mutations or of intelligent design. Thus, if evolution theory is ever falsified, the design/creation alternative will be the only viable theory. Pessimistic meta-induction (Laudan, 1981) argues that we have no reason for thinking that any of our current theories are true since the history of science is a "graveyard" of once empirically successful theories. Thus, based on this argument, it is highly possible that evolution theory is also false. It appears that intelligent design supporters have failed to capitalize on his argument.

A brief introduction to the structure of scientific theories

Since the time man started to organize acquired knowledge about the world in a body we nowadays call science, there has been a continuous process of developing and evaluating *explanatory hypotheses*. An explanatory hypothesis can be a statement, or a logical argument that is intended to offer an explanation of a phenomenon. In modern science, an explanatory hypothesis must also be subject to falsification by experiment. A well-defined process for the development and testing of explanatory hypotheses is known as the *scientific method*. This method relies on experiment and testing for eliminating false hypotheses.

So what distinguishes a scientific hypothesis from a non-scientific one? For example, it is easy to understand why hypotheses that involve supernatural causes are not allowed in science and thus are non-scientific; it is because they cannot be falsified experimentally. For example, an explanatory hypothesis asserting that life is the result of intelligent design cannot be falsified experimentally. There is simply no way to device an experiment so that such a hypothesis is tested in a direct way. Neither such hypothesis generates any testable predictions that can be falsified by experiment, at least not any that we understand. On the contrary, the hypothesis that all swans are white is falsifiable and it was actually falsified by detecting a black swan. It is important to understand that failure to detect a black swan does not prove that the hypothesis that all swans are white is true and any such claim is an *argument from ignorance*. What matters in science is whether a hypothesis is falsifiable and a failure to falsify a hypothesis cannot be used to assert its truth.

Principles, postulates and doctrines

At the core of contemporary science theories, one finds the notion of a principle. A principle can be a fundamental law, an assumption, a postulate or even a doctrine. A postulate – also called an *axiom* – is a proposition that is assumed to be true without any evidence. A postulate can be falsifiable or non-falsifiable. A doctrine can be a philosophical position or even a system of belief. Contemporary science has increased the flexibility of developing theories by effectively utilizing these concepts. Modern theories are not based on just one hypothesis. They usually involve several principles and generate predictions in the form of falsifiable hypotheses. The principles and predictions of modern theories are tested by scientists employing the scientific method. For example, the Higgs boson and associate field is a hypothesis generated by the *standard model of particle physics* and is falsifiable. Scientists using particle accelerators for testing this hypothesis follow the scientific method.

It is important to understand that not all the principles used in a theory must be falsifiable as long as the predictions of the theory that are tied to these principles are falsifiable. One of the principles of Einstein's special relativity is that space and time form a four-dimensional continuum, called *spacetime*. This principle is essentially a postulate and physicists refer to it as one of the postulates of special relativity. It is assumed to be true. One cannot prove or falsify this postulate because spacetime in that particular multidimensional form is not directly observable. But in conjunction with other postulates and empirical principles, Einstein's theory of special relativity in its modern formulation generates many testable predictions that have been verified in laboratories numerous times. So far, none of the predictions made by special relativity have been falsified in a laboratory experiment, although scientists have targeted them intensely and the theory is accepted as a valid, high precision model in its domain of application. Those who believe that there is a conspiracy not to prove special relativity false are flat wrong. Scientists know that progress is made when theories are falsified and replaced by alternatives that make better testable predictions. People are also incorrect when they believe that a theory, like special relativity for example, can be proven true. Theories can only be corroborated or falsified by experimentation but they can never be proven true.

The use of non-falsifiable postulates by modern science

Scientific theories do not contain only falsifiable principles or postulates. For example, although two of the four postulates of special relativity are falsifiable, the principle about the four-dimensionality of spacetime is not. The falsifiable principles are that the speed of light is constant in any inertial reference frame — although this has been debated for too long whether it is the case— and that the laws of physics are the same in all inertial reference frames. There is a fourth principle about the existence of globally inertial reference frames and that has the implication that the geometry of spacetime is flat in that theory's application domain. It is debatable whether this last principle should be considered in the same sense as a falsifiable principle or just as a useful mathematical convention for the formulation of the theory, but an in-depth discussion of such conventionality issues lies outside of the scope of this book. However, the principle about the four-dimensionality of spacetime has direct implications on measurements made locally by various observers that move relatively to other observers. More importantly, the choice of spacetime structure commits to certain metaphysics about physical reality. A four-dimensional spacetime with flat geometry is a different world from one with an absolute three-dimensional space plus universal time, in the way Newton thought of the spacetime structure. In the former physical reality, simultaneity is relative to a reference frame and there is no objective here and now, whereas in the latter physical reality there is objectivity about location and time. These are two distinct types of physical reality and they commit to a different ontology. For example, in a four-dimensional reality, an argument can be made against the existence of three-dimensional substances, a concept that is considered common sense in three-dimensional physical reality. (These issues are discussed in more detail in Chapters One and Four).

Theories can only be corroborated

It is also important to understand the fine distinction between theory proof and theory corroboration. Let us consider a simple example in order to understand why a theory, or even a single falsifiable hypothesis, can never be proven by experimentation but only corroborated. Consider a door leading to a room and Jack's theory is that "the door is always locked". This theory has just one hypothesis that it is falsifiable. The experiment one can do to falsify the theory is simply to try to open the door. If no key is needed to open it, then the theory is falsified. It only takes one experimental observation contrary to the theory prediction to falsify it. But what about proving the theory?

Let us assume that no matter how many times Jack tries, the door cannot be opened without a key, not even after 1,000 or 10,000 times. Does this mean the theory is true? The answer is no, because it may be that the next time that one tries to open the door it unlocks for some unknown reason. The main point is that the theory claims that the door is "always locked". From a finite, even large number of observations, Jack may infer that the door is always locked with high probability but he can never be 100% certain. This type of inductive reasoning, meaning the formation of a general rule from a number of observations, cannot provide the absolute certainty needed for proving any theory true. Such reasoning can provide only a probability about the door being locked the next time that Jack tries to open it. The probability may be high and close to 100% but it can never become exactly equal to 100%. This is true even if the number of trials becomes extremely large. In that case the probability will approach 100% asymptotically at the limit. If an event has probability equal to 100%, then this is a certain event, such as an event that has already occurred.

Jack may raise his theory to the status of a principle and use it in another theory with other hypotheses or principles.

However, observations, no matter how many and how precise, cannot constitute proof that a given theory is true or that it is a law of nature. The law can only be corroborated by observation but never proven to be true. This is related to the well-known problem of induction.

Intelligent design as part of a scientific theory

As I have already mentioned, intelligent design, in the form of an explanatory hypothesis, is neither a rule obtained by inductive reasoning, nor a falsifiable proposition,. However, as in the case of Einstein's special theory of relativity and the use of the non-falsifiable postulate of four-dimensional spacetime, a variation of the intelligent design hypothesis could be used as a postulate in a theory and not as a direct, bold explanatory hypothesis. Along with other principles and postulates, this new theory could generate new and testable predictions that are related directly to that postulate and are falsifiable. But that would not corroborate the truth or falsity of the new postulate, but only that of the theory. The theory of evolution also has several principles, some empirical and some in the form of a postulate. One of the principles involved is natural selection and although there is a dispute whether it is a falsifiable one, the resulting theory produces testable predictions and it is therefore falsifiable.

A false claim of some intelligent design proponents is their assertion that the principle of natural selection cannot be falsified and hence the theory of evolution is not scientific. Again, this may be true in the sense of direct falsification but the modern theory of evolution is falsifiable and this is what matters. By attacking just one principle of a theory it reminds of the thousands of mathematicians who spent their whole life trying to prove or disprove Euclid's fifth postulate. In a nutshell, Euclidean geometry has five axioms, and every theorem of that geometry can be derived from them. Although the first four axioms can be assumed in a certain sense to be self-evident, the fifth axiom, that essentially claims that two parallel lines meet only at infinity, is not self-evident or has not been assumed to be such by mathematicians. Many have tried to either deduce the fifth axiom from the other four or prove that it is either true or false, but such exercises have proven futile. However, a few bright mathematicians – amongst the first was Gauss – understood that what was important was not to try to prove the fifth axiom true or false. They took advantage of the fact that it was a weak axiom and were able to come up with two other consistent systems of geometry, known as elliptical and hyperbolic. This was achieved by replacing the weak fifth axiom and some of the others with other propositions such as, for example, that any two parallel lines never meet or that they always meet. In a similar way, intelligent design proponents may find it more profitable to use their hypothesis or a variation of it in some suitable form in a new and falsifiable theory, rather than trying to prove their opponent's theory false, which is something that possibly cannot be done.

A theory that includes an appropriate variation of the intelligent design hypothesis as one of its postulates will have value provided that such a theory can generate testable predictions directly related to that postulate that are falsifiable. But that may disappoint some hard core intelligent design supporters who seek an acceptance of their bold hypothesis without questioning it. Acceptance of any non-falsifiable hypothesis as true or as a law of nature is impossible within the constraints and limitations imposed by modern science. One could better use such constraints and limitations to one's own advantage but this requires hard work and innovation. It goes beyond just a senseless debate, a heated discussion or dogmatic stand and moves into the realms of productive work with an aim of advancing science and human understanding. This is the aim of this book.

Therefore, I argue that it may be possible to convince the scientific community to investigate alternative theories that incorporate a suitable variation of the intelligent design hypothesis as a postulate. This could be achieved after demonstrating, or actually reminding people I should say, of the foundational problems of science. Then, go one step further and show how a new theory can offer an alternative view of our world and make new predictions that are testable and falsifiable. This is the strategy followed in this book. But in the process of achieving this goal, it will be necessary to consider intelligent design as part of a postulate much broader in scope, called an *intelligent interaction.* In a nutshell, this new postulate asserts that the universe, the laws of physics and life exist because there is intelligent interaction between a universal mechanism and the constituents of the phenomena. This mechanism is part of the physical universe but it is not directly observable, in the same way that spacetime is not directly observable.

The idea of an interacting cause that brings about all phenomena is not new. As we shall see later in the book, Cartesian philosophers proposed a similar provocative but, in my opinion, advanced postulate during the 17th century disguised as the *Action of God.* Such a postulate, in a modern and suitable form, can lead to a theory that can provide an alternative foundation of science. I will thus attempt to show in this book that it is possible to have a foundation of science based on the postulate of intelligent interaction. The consequences from such an approach are surprising and have important implications about the nature of our physical reality.

In order to show how the postulate of intelligent interaction can lead to an alternative foundation of science, I will take you on a journey in time and space that starts at Elea, at around 550 BC. Elea was a small ancient Greek colony in the south of Italy. It is where the strongest blow to common scientific wisdom was ever delivered against any attempts to understand the world through observation, experimentation and inductive reasoning by a group of poets and philosophers. I will attempt to clarify the meaning of their arguments and their motivation, which scientists and philosophers since the time of Plato and Aristotle have often misunderstood, or even tried to obscure. I will then proceed to reconcile the Eleatic arguments with the ideas expressed by Cartesian philosophers of the 17th century, discuss how Descartes, Newton and Leibniz dealt with them and then discuss some of the foundational weaknesses of modern scientific theories, special relativity and quantum mechanics.

I will even go one step further: I will present a theory of mechanics in place of Newton's that is compatible with the postulate of intelligent interaction. I will argue that a mechanical and material world does not preclude an intelligent and constantly interacting cause that brings about the phenomena, but that such cause is actually necessary for the world to exist. I will show a few examples where observations are in conflict with theory, embraced models, and common scientific wisdom.

There will always be those who will insist that explanatory hypotheses that make direct ontological claims, meaning hypotheses about the true features of reality in a metaphysical sense, should not be part of science but fall in the realms of naïve realism. However, it is hard to argue that modern scientific theories do not commit to a particular ontology. The eclectic stand of ignoring any ontological commitments, especially where they come in conflict with logic, common sense and experience, while at the same time embracing a theory only from an epistemological stand, is not good service to science and humanity. Also, it undermines the desire of man to understand the nature of physical reality and the origins of life. The objective of this book is not, however, to try to prove any metaphysics. The sole objective is to show how the postulate of intelligent interaction can be used in a theory that provides an alternative view of our world and an alternative foundation of science. This new alternative may serve to balance current paradigms and their impact on the cultivation of human values.

Towards an alternative foundation of science

The approach taken in this book is thus one of developing an alternative theory in which intelligent design is implied by one of its postulates, called intelligent interaction. This is a scientific theory in a sense that it provides new testable predictions. Experimental success can only corroborate the truth of the theory but never prove it, as discussed earlier. Any gains from corroboration will be limited to an acceptance of intelligent interaction as one of the principles of a new theory, and as a consequence the plausibility of intelligent design. I would argue that although this may not satisfy some proponents of intelligent design it would nevertheless be a decisive first step and a win against scientific materialism.

One might wonder what the purpose of this effort is. What difference would it make if we knew that our world and human beings are the outcome of a random process or of intelligent interaction? How could that affect our everyday lives and future? I will also try to answer these important questions in the conclusion at the end of the book. I have already mentioned the need for a balance to current paradigm and their impact on the cultivation of human values. However, the answers will only carry some value for those that will read the material and follow the path of the book

A Note to the Reader

This book contains a fair amount of formulae. This is unavoidable since science is almost synonymous to mathematics, or said in a poetic way, *math is the language of science*. If you are not familiar with the simple mathematics presented in the book you can skip any derivations and formulas. After each formula or derivation, an effort is made to offer an explanation in words that provides the essence of what was accomplished with symbols, its meaning and use.

Section I – PARADOXES, IMPOSSIBILITIES AND INCONSISTENCIES

CHAPTER ONE: THE IMPOSSIBILITY OF MOTION

It is hard to conceive life without motion, a world where nothing moves. Life in modern societies is almost synonymous to motion, such as taking the bus to work or going for a walk, driving a car, flying to another country on a vacation or even going around earth in orbit. Since antiquity, motion has been assumed to be a primitive notion. Heraclitus (540 – 475 B.C.E) with his doctrine that "all is flux" made an early attempt to raise motion to the status of a universal principle that could serve as the starting point for understanding the phenomena of our world.

Yet, the Eleatic philosophers were not convinced. These philosophers of the 5th Century BC were well-known for their provocative views about physical reality, which came into direct opposition with Heraclitus' doctrine of universal flux. Monism was the central doctrine of the Eleatic school of philosophy founded by Xenophanes (570 – 475 BCE) and was based on the principle that "what exists, can only be one and motionless". Parmenides (b. 510 BCE), who was a student of Xenophanes, is the best known of the Eleatic philosophers and the one who amended the original doctrine as follows: "what exists, can only be one, motionless,[3] indestructible, immutable, finite and indivisible" (Mostratou, 1948). In this way, Parmenides made a clear distinction between Eleatic philosophy and other schools of philosophy of his time, and in particular, took an opposing stand against Heraclitus and his doctrine of universal flux. According to Parmenides, there is no motion, no plurality and no void, and that comprises the basic thesis of the doctrine of monism while it clearly differentiates it from other doctrines.

---

3   I have used the word motionless rather than immovable in translating the Greek word ακίνητος.

At first, monism sounded like an incomprehensible or even a naïve stand. However, the logical argument Parmenides put forward was disturbing, to say the least.. In Περί Φύσης (About Nature), Parmenides argued that since all things are what they are and not something else, then it is impossible for one particular thing to have and to not have some feature, and as a result a change from having one feature to not having it is also impossible. So according to Parmenides, change, in the sense of changing features, is apparent and we must differentiate between what we perceive and the reality of things. In essence, Parmenides directly challenged the reality of the physical world, as we perceive it. This was also an early argument in favor of *rationalism,* the epistemological theory that true knowledge about the world can only be achieved *a priori* through pure reason. At the other extreme is *empiricism,* the epistemological theory that knowledge can only be acquired *a posteriori* through experience.

Pluralism and discrete atomism were attempts to rebut the Eleatic claims and to rescue the reality of the physical world (Guthrie, 1965). Empedocles (d. 433 BCE) and his student Anaxagoras (500 – 428 BCE) reacted swiftly to the Eleatic views. The central doctrine developed by the pluralists, as these philosophers came to be known, was that "a beginning requires more than one". Furthermore, they claimed that all elements in the world are infinitely divisible. The pluralists opposed the Eleatic notions that there is no motion and no change. Anaxagoras agreed with Parmenides that there can be no destruction of elements and that there is no void.

The notion of *void* was incomprehensible for the Eleatics since for them it was associated with nothingness and could not exist by definition. On the other hand, for the pluralists, the existence of void was not ruled out *a priori* but *a posteriori* because of the requirement of infinite divisibility. If our world is material, then according to the pluralists there could be no void and everything is infinitely divisible. Therefore, for both the Eleatics and the pluralists, the world is a plenum.

Pluralism offered hope of a rational rebuttal of Eleaticism until Zeno of Elea, a student of Parmenides, presented his logical arguments against plurality, which were also known as Zeno's paradoxes or plurality paradoxes. It soon became evident that pluralism was not an adequate challenge to monism. In turn, Leucippus (c. 450 BCE) introduced the concept of causality in an effort to rebut the Eleatic claims after the failure of pluralism. "Nothing happens without a cause but everything happens because of reason and necessity", is the only phrase preserved from Leucippus. Democritus (460 – 370 BCE), a student of Leucippus, developed the theory of discrete atomism by being the first to introduce the notion of an atom and motion in a void. In essence, Democritus agreed with Anaxagoras about plurality, motion and change, but disagreed about infinite divisibility and the non-existence of void. Thus, the atom was postulated to be an indestructible and indivisible unit that can move in voids. The key point to understand is that the atomists tried to restrict monism within the notion of a single unit or atom by saying that there is no motion, no void and that the atom is indestructible, indivisible and does not change. Barnes (1979) argues that "atoms originated in Elea", a figure of speech of course, as the notion of an atom is incompatible with the Eleatic doctrines, which oppose the existence of discrete structures and motion in a void. The truth is that atoms originated as a reaction to Eleaticism. Nevertheless, the concept of the atom was an ingenious but unsuccessful attempt to rebut the Eleatic claims. As it turned out, atomism was a much easier target for Zeno and his paradoxes than pluralism had been.

Zeno of Elea and the Paradoxes of Motion

Zeno of Elea (c. 450 BCE) was a student of Parmenides, who later in his life moved to Athens and served as a mentor of some influential men, including Pericles, a proponent of a democratic political system. Zeno is best known for his paradoxes of motion, a set of logical arguments concluding that motion is impossible in either continuous or discrete space. Three of these paradoxes are known as the *dichotomy, Achilles and the tortoise, and the arrow,* although more than 42 paradoxes are attributed to Zeno. There have been a great number of interpretations and proposed solutions to these paradoxes throughout the ages. It is likely that most of the proposed naïve solutions to Zeno's paradoxes were due to a lack of understanding of the philosophical framework within which the paradoxes were developed and, perhaps, an effort to identify all of Eleatic philosophy within the specific argument structure of these paradoxes.[4]

A frequent naïve thesis is that Zeno attempted to prove that the perceived phenomenon of motion is not possible. What Zeno was in fact trying to prove, is that what we perceive as motion cannot be a change in the location of matter in space, as it is commonly perceived, since motion is impossible in either a plenum or vacuum. There is a fine and important distinction between these two different interpretations that can lead to substantially different approaches in searching for possible solutions to the paradoxes.

Zeno essentially argued then that what we perceive as motion of objects in three-dimensional space is in fact something else, because motion, in the sense of physical movement, is impossible according to the Eleatic doctrines. This impossibility arises at the highest level after the realization that motion in a plenum, an infinitely divisible space according to the pluralists, is impossible, because there is no place for things to move. Additionally, motion in the void of the atomists is also impossible because things cannot move in the absence of a medium. Therefore, what have all the naïve

---

4   The paradoxes survived in time, because on many occasions Zeno presented his arguments in front of a live audience. On one of these occasions, during Zeno's efforts to convince his audience that motion is impossible, the philosopher Antiphon rose from his seat and walked in front of Zeno, in silent protest against his arguments. Despite the protest by the philosopher, by the end of the presentation, there was a clear consensus within the audience that there is no such thing as motion (Mostratou 1948, p. 76).

solutions to Zeno's paradox accomplished so far? Obviously nothing was achieved, other than to reveal the magnitude of the problem. Even in the case of the dichotomy paradox, if Zeno in fact ever expressed it in the particular form it is often stated, the proposed solutions are not convincing at all. My attempt to illustrate the reasons for this follows.

Zeno's dichotomy paradox (popular interpretation): motion is impossible because in order to travel a given distance, an object must arrive at the middle of that distance first and just before that, the object must arrive at the middle of the first half interval and so on, ad infinitum.

The dichotomy paradox has its roots in a version of Zeno's plurality paradox stating that "many beings do not exist". This latter paradox is also called an antinomy in logic, a term that describes a pair of conclusions that are contradictory. Zeno asserted along these lines that, "if there are many things, it is necessary that they are as many as they are, and neither more nor less than that. But if they are as many as they are, they will be limited. If there are many things, the things that are, are unlimited; for there are always others between the things that are, and again others between them." And thus the things that are, are unlimited. (Kirk et. al, 1983, p. 266) Therefore, the proposition "many things exist" leads to a pair of contradicting conclusions, an antinomy.

The dichotomy paradox was an attempt by Zeno to illustrate a process that generates an infinite number of intervals to transverse from a finite distance given that plurality is assumed. This is what is referred to in modern literature as a supertask. The important point is that the dichotomy paradox is a statement about the impossibility of motion in a plenum. The particular form, in which it is often expressed, is merely an illustration of the nature of the problem one is required to solve in order to prove that physical motion is possible in a plenum. Solutions to the paradox based on infinite series convergence rest on a number of assumptions designed to only guarantee formal mathematical consistency and not to prove that motion in a plenum is possible.

A naive claim of a solution to the dichotomy paradox is that although an interval can be divided into infinite sub-intervals, the infinite sum converges to a finite number. For the dichotomy paradox, this is illustrated in figure 1-1, where a unit interval between points S and E is dissected into successive sub-intervals. The first interval is from point E to point α_1 and has a length equal to ½, the next is from α_1 to α_2 and has a length equal to half the preceding interval, or equal to ¼, and so on.

There are at least two interpretations of the dichotomy paradox. The first is based on motion starting at point E and reaching the first point, α_1. Then, the moving body must travel to the second point and so on, ad infinitum. Since there are infinite points to cross, Zeno claimed the object would never reach the end, point S.

The second interpretation, which in my view is a lot more interesting, has the moving body starting at point S. Since there are infinite points between point S and point α_1, the moving body will never make it to the first half of the interval. However, since given any point close to S, there will always be another point closer to it and due to the infinite divisibility property assumed, motion cannot even commence.

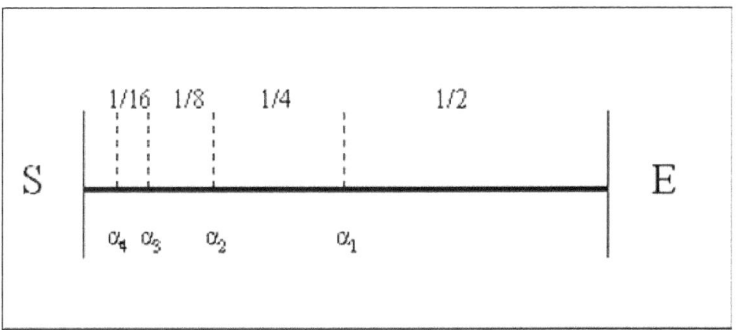

Figure 1-1: Infinite number of sub-intervals in the Dichotomy Paradox

The series of the infinite sum of the sub-intervals for a unit distance ES in the dichotomy paradox is given by:

$$\sum_{n=1}^{\infty} \frac{1}{2^n}$$

A naïve mathematical claim is that "the dichotomy paradox is unsound because the above infinite series converges to 1", which is the length of the interval to travel. But such a claim involves several definitions and assumptions about sequence and series convergence designed just to guarantee consistency. In a nutshell, an infinite series converges to a real number σ if, and only if, the sequence of partial sums σ_n converges to that real number σ: In formal mathematical notation, this is expressed as follows:

$$\sum_{n=1}^{\infty} a_n = \sigma \Leftrightarrow \lim \sigma_n = \lim(a_1 + a_2 + \cdots + a_n) \equiv \lim \sum_{k=1}^{n} a_k = \sigma$$

In turn, a sequence σ_n converges to a real number σ by definition if:

$$\lim \sigma_n = \sigma \Leftrightarrow \lim(\sigma_n - \sigma) \to 0$$

Furthermore, the definition for the convergence of a sequence, in this case $(\sigma_n - \sigma)$, to zero is the following:

$$(\sigma_n - \sigma) \to 0 \Leftrightarrow \forall \varepsilon > 0 \quad \exists n_0 = n_0(\varepsilon) : |\sigma_n - \sigma| < \varepsilon \quad \forall n \geq n_0$$

This last definition states that a sequence $(\sigma_n - \sigma)$, converges to zero if, and only if, for every ε greater than zero, there exists an index n_0, depending in general on ε, such that the absolute value of the sequence is less than ε, for every value of n greater than n_0. In the case of an infinite series arising from infinite divisibility of a finite quantity, the convergence proof rests on such definitions. The definitions obviously result from a consistent mathematical framework for assuring that the finite mathematical quantity, such as the length of a distance to transverse, never becomes infinite by sub-dividing it into infinite segments and then adding them back. In other words, such definitions guarantee that distance remains invariant when it is divided into infinite pieces that are then summed up. *But such definitions when used to corroborate physical motion assume the possibility of motion in the first place.* Is this mathematical solution in any way relevant to the problem posed by Zeno's dichotomy paradox, which involves physical motion in an infinitely divisible plenum? Hardly so, since these mathematical definitions fail to resolve the higher-level issue of the possibility of motion in a plenum or vacuum.

Modern mathematics has made a return to classical proofs of convergence using infinitesimals, by postulating the existence of pseudoreal numbers, which belong to a nonstandard universe, and such numbers are infinitesimal (McLaughlin and Miller, 1992). Such solutions about infinite series convergence also remain, however, in the realm of formal mathematical consistency. They do little to address the fundamental higher-level problem of physical motion in a plenum, or vacuum, and the apparent distinction Parmenides and Zeno made between how we perceive the phenomenon of motion and the reality of the world.

There are even further complications arising from the realization that in the dichotomy or Achilles and the tortoise paradox, one is faced with an infinite machine supertask (Benacerraf, 1962). The infinite machine supertask concept can lead to contradictions in the framework of classical physics but an in-depth analysis is beyond the scope of this book (Svozil, 1995). It may suffice to say that it is impossible to develop an algorithm that will divide a given distance into infinite sub-intervals while at the same time avoid an infinite loop. Infinite series convergence is only asymptotic, meaning that the final destination is never reached exactly. In another sense, if at every instance when a moving particle passes the middle of the remaining distance from point E to point S some type of information had to be recorded in some electronic medium, it would require infinite memory for doing so. This is a problematic realization because an infinite amount of space-energy is required to record actions that arise in finite space by finite energy.

What about the possibility of motion in a void, assumed by the early atomists and modern science? Zeno objected to such a notion with his arrow paradox.

The Arrow paradox: An arrow at rest is in a place equal to its own size, and at any instant or time interval during its flight the arrow is also in a place equal its own size, therefore the arrow is always at rest.

With his arrow paradox, Zeno questions the phenomenology of motion, in the sense that there is nothing deduced directly from the phenomena that motion is taking place or that is even possible. He also uses the paradox to challenge the notion of discrete space and time. Zeno argues that something cannot move in a void since the change in location form point E to point S, when there is nothing in between the two points, is impossible according to the Eleatic doctrines put forward by Parmenides. Specifically, it is impossible for one particular thing both to have and to not have some feature, in this case the void, and as a result a change from having one feature, like the state of void, to not having it and becoming part of an arrow, which is also impossible.

There has not been a convincing solution to Zeno's paradoxes so far. Barnes (1979, pp. 231 – 294) offers an extensive analysis of the paradoxes just to conclude that he cannot show that they are unsound because he cannot prove the fundamental premise of Zeno that 'nothing can perform infinitely many tasks' to be false. Remember that in the dichotomy paradox there are infinite points to pass. If each passing of a point is considered as some type of a task, then there are infinite tasks to perform. Despite the conclusions of many researchers that there is no convincing solution to the paradoxes not involving some kind of fundamental assumption about the nature of motion, the objections of Zeno and the Eleatics were dismissed throughout the ages by mainstream science with no clear justification. Whenever someone reminded them of the paradoxes, scientists would just reply with naïve arguments or even aphorisms. Yet, the foundations of classical mechanics, which include kinematics and dynamics, took the concept of motion as self-evident and Galileo's law of inertia was based on the notion of motion in a vacuum. As we shall see in Chapter Three, Newton's second law, also known as the force law, was based on the time rate of change of a quantity he called motion and his law of universal gravitation was based on forces acting on masses across an intervening void. In other words, the foundations of classical physics were based on notions the Eleatics argued, successfully according to some, that they are naïve and paradoxical, to say the least.

The Eleatic philosophers tried to emphasize the distinction between phenomena and reality, the latter being the totality of what truly exists, in a metaphysical sense. From the phenomenon of motion, for example, we cannot find an empirical explanation as to why and how such change is physically possible, either in a plenum or vacuum, based on traditional conceptions of space and time. Instead, the phenomenon leads to logical contradictions when considered in the context of naïve notions of space and time, as Zeno asserted with his paradoxes.

Aristotle, the original inventor of *Categorical Logic,* tried to find a rational explanation for Zeno's paradoxes. His attempt was based on identifying possible hidden premises that Zeno used and then proving that at least one of them was false in order to declare his arguments unsound. Aristotle argued in Physics 236a15 that Zeno in his dichotomy paradox should also divide the time taken to reach a certain point, in addition to dividing the distance. Aristotle then claimed that if time can be divided geometrically, then that should also be the case physically. But, he argued, since time cannot be divided because motion would be discontinuous, then also distance cannot be divided. In essence, Aristotle denied the existence of static instants in time (Shamsi 1994). In addition he claimed that since motion is a phenomenon of experience, Zeno's argument is a dialectic. A detailed discussion on how Aristotle attempted to cope with Zeno's paradoxes can be found in Barnes (1979, pp. 261 – 295). It appears though that Aristotle avoided dealing with the essence of the paradoxes, which is the impossibility of motion in a plenum or vacuum although he was an advocate of the plenum thesis. Nevertheless, Zeno never insisted time or space could actually be divided into infinite points but only that an infinite number of points exist naturally in a plenum in the form of a collection defining a certain path. The dichotomy process he used was just an illustration of the problem one is faced with when trying to model the phenomenon of motion in a three-dimensional universe where the only events that exist are in the past and present time. In that sense, he did not have to divide time or even use it at all in the process. Zeno did not insist that motion along a given path must take infinite time but that it is impossible to even occur in a plenum of vacuum, the part of the argument that Aristotle seems to have avoided to deal with.

Motion and change in modern science

The Eleatic doctrines and especially Zeno's paradoxes challenged naïve perceptions of space and time by showing that they lead to logical contradictions. These challenges were either neglected or in many cases misunderstood by some of the pioneers of classical mechanics, other than with the exception of some Cartesian philosophers as we shall see later on in the book. Even in the latter case, there was a definite dose of obscurantism. These philosophers never made the deeper reasons for their philosophical arguments and their conclusions regarding physical reality explicit. Few can get to the bottom of their complex arguments and understand what drove them to such a level of complication. We will discuss how Descartes approached the problem of motion in more detail in the next chapter. A number of contemporary scientists and philosophers understood the meaning and metaphysical implications of the Eleatic doctrines and their arguments well, but they also avoided discussing their importance publicly or in their writings. After all, science would be humiliated to admit that a few ancient paradoxes are still unsolved.

Relativity, and in particular special relativity, changed the classical views of a world where substances with intrinsic properties, such as a body with extension, move in three-dimensional space where there is objective simultaneity of events. However, this change came at a high price that most scientists are still not convinced has to be paid. Most accept relativity theory as an accurate mathematical model in an instrumentalist sense but not its metaphysical commitments in a realist sense. As a result, they doubt that relativity is a true description of physical reality.

Specifically, relativity commits to a tensless theory of time and existence and to events instead of substances (Esfeld, 2009). This is a drastic departure from the metaphysical commitments of classical physics. According to relativity, simultaneity is only relative to a reference frame and is not objective. As a result, an event that it is the past of an observer can be in the future of another observer depending on their relative state of motion. Furthermore, everything that exists in the past, present or future, exists in a four-dimensional block universe.

The objects in the block universe are different from the objects of classical mechanics and of a three-dimensional universe. Relativity theory objects do not only have geometric parts, they also have temporal parts. Their extension is not fixed, as in the case of rigid three-dimensional objects but it is dependent on the reference frame of the observer. In relativity theory, physical objects are continuous sequences of spacetime points, also known as processes. According to this radical departure from classical mechanics, motion in a block universe can be defined as a continuous sequence of spacetime points that have similar physical content. Change is a continuous sequence of spacetime points with different physical content.

Let us try to understand the implication of the concepts of motion and change in a block universe. In Zeno's dichotomy paradox (see Figure 1-1), we considered an object that moves along a certain distance in three-dimensional space from point S to point E. Zeno tells us that this is impossible either in a plenum or vacuum. However, if we could merge the notions of space, objects and time, the paradox no longer applies in this new universe. In this universe, there is space-time and there are processes. In order to accomplish this, we must consider space and time as a four-dimensional continuum. The consequence is that the concepts of objective "now" and objective "here" are abolished together with the notions of substance. It is important to understand that, in this way, we essentially submit to Zeno's doctrines in order to resolve his paradoxes. The universe of relativity is Parmenidean, in the sense of a block universe; it is immovable and timeless. Motion and change are defined as sequences of spacetime points that are already here and there. In three-dimensional reference frames, these sequences give rise to phenomena such as objects and their motions. Everything that happened in the past or will happen in the future is already there as a continuous collection of spacetime points.

There is another way to look at the solution to Zeno's paradoxes from the point of view of relativity and a block universe. An object in three-dimensional space starts at point S at time t_1 and arrives at point E at time t_2. Zeno tells that this is impossible and we are observing a phenomenon that cannot physically occur in three-dimensional space and absolute time. Relativity comes in and tells us that the object is a continuous collection of spacetime points and also $\{E, t_1\}$ and $\{S, t_2\}$, the start and the end, are spacetime points that belong to that sequence. The collection of all the spacetime points defines both the object and its motion. The object starts and finishes its motion in three-dimensional space because this collection of points and associated process are part of spacetime. It is not a problem if these points are infinite as long as they are part of continuous spacetime. This is the relativity answer to Zeno. In essence, the relativity view abolishes the notion that there are three-dimensional objects that move in space and replaces it with the notion of processes in space-time. For many, this is a high price to pay.

There are those who will argue that the notion of three-dimensional objects with extension is compatible with special relativity. This view leads to inconsistencies, like the bug-rivet paradox, as we shall see in Chapter Four. One is forced to accept only four-dimensional objects in special relativity, so that they retain their physical shape relative to all inertial reference frames. Thus, objects must have temporal parts in addition to physical extension to avoid inconsistencies. However, if there is actually a preferred reference frame it means special relativity is not a true theory of nature. The ultimate consistency of special relativity depends on whether its axioms are true. Those that insist that special relativity is an inconsistent theory because the principle of the constancy of the speed of light and the principle of relativity combined lead to contradictions, often do so because they commit to the existence of three-dimensional objects. Actually, Albert Einstein was correct to include the constancy of the speed of light postulate because it made the theory consistent. It is either another matter whether special relativity is a true theory of physical reality or that it is just an instrument for making predictions. If there is an unobservable preferred reference frame which serves as the arena in which three-dimensional objects take part in events, then, in that case, relativity is just a useful model but Zeno paradoxes are not resolved given the new metaphysical implications of a preferred reference frame.

Should we then reject three-dimensional substances and a preferred reference frame and commit to four-dimensional objects, constancy of speed of light and a block universe? Isn't the price paid too high? I cannot answer these deeper questions. My main objective was to show how a solution that modern science offers to some old paradoxes has just managed to elevate complexity to a new and higher level and replace old foundational problems with new ones.

A disturbing message with elusive solutions

The Eleatic philosophy and specifically Zeno's paradoxes carried a disturbing message to those who hoped that the world could be explained by observing and analyzing the phenomena. Zeno of Elea argued that what we perceive as motion is not what we think it is, because what we perceive as happening cannot happen if one employs common sense notions of space and time. Many philosophers and scientists during those early times started having serious doubts about man being able to ever explain anything at all besides developing some models that simply describe what happens, but not why or how it happens. In the *Republic,* Plato used the *Allegory of the Cave* to give a possible explanation as to why his call to "save the phenomena" had failed. Science is trapped in a cave, Plato thought, which is a world of phenomena and experimental evidence. This makes it difficult for science to accept anything outside of what exists in the cave simply because science and the scientific method can only deal with what is inside the cave.

Modern science has made a dramatic shift away from traditional notions of classical physics in an effort to make better and more accurate predictions about the phenomena. The metaphysical commitments of modern scientific theories are hard to accept especially when it comes to the notion of four-dimensionalism in relativity. People just find it difficult to commit to notions such as that the past, present and future all intermingle in a universe where objects have temporal parts. Are these naïve attempts to resolve paradoxes or a necessary and decisive step towards a better understanding of the world? Have these new concepts removed ambiguity or created more? Why is it that these theories make more accurate predictions in exchange for refuting some notions that were perceived as common sense for a long time? These are questions that remain unanswered. It appears that a new and more sophisticated level of complexity and ambiguity has emerged and that may be an indication that we ought to revise our basic assumptions about the nature of reality if we are to make progress towards a better understanding of the world.

CHAPTER TWO: THE IMPOSSIBILITY OF MECHANISM AND DYNAMISM

After almost two millennia from the time of Zeno, the reasons as to why ancient Greek scientists failed to save the phenomena were almost forgotten or even undermined in a way. A new era started when scholastic philosophy of the medieval period gave way to natural and experimental philosophy and that shift paved the way towards the developments of the body of knowledge we nowadays call physics. Thomas Hobbes (1588 – 1679) in Britain, Rene Descartes (1596 – 1650) and Pierre Gassendi (1592 – 1655) in France and Galileo Galilei (1564 – 1642) in Italy were contemporaries that, amongst others that are less known but maybe as equally important, proposed a purely mechanistic view of the world. This new view was to serve as an alternative to medieval scholastic philosophy and the employment of final causes.

For Hobbes, everything is material including human beings who are nothing more than complicated machines and their actions can be totally explained by studying the interactions of matter. For Descartes, the tools for achieving knowledge are mathematics, pure reason and the use of indubitable propositions, whose truth cannot be doubted. His natural philosophy was purely mechanistic and, according to it, all physical phenomena, apart from the human mind can be explained based on interactions of matter. Gassendi adhered to a theory similar to ancient Greek atomism that could offer a mechanistic view of the world, but not of the human mind. For Galileo, experimentation ought to be a fundamental part of science and thus he is considered the father of modern physics.

The important thing to understand is that these early developments were attempts to establish a foundational framework that would support the development of science on the basis of the autonomy of the world and in the sense of abolishing the need of supernatural causes for explaining the phenomena. It was during that period that the foundations of methodological naturalism were set.

Mechanism and corpuscularianism

The works of Rene Descartes and in particular his book *The World* (Le Monde) had a profound impact on Western Civilization and on the development of modern science, despite the fact that most of the credit for the progress made was later assigned to Galileo and Isaac Newton. Descartes' mechanical philosophy was based on a model of the universe consisting of "corpuscles" of matter. He argued that matter was in constant motion and formed a plenum in which no void could exist. In a sense, the model of Descartes incorporated Heraclitus' doctrine of universal flux and pluralism, the theory developed by Empedocles as a response to the Eleatic philosophy, as discussed in Chapter One. According to Descartes, there is no void; matter is infinitely divisible and forms a plenum. Motion occurs due to matter moving in vortices and pushing other matter.

Descartes spent a great deal of effort in his treatise Mediations to denounce discrete atomism and the notion of atoms moving in voids. Thus, he did not agree with Gassendi that atomism could serve as a foundation of a contemporary science although they both agreed that the human mind could not be explained mechanically. However, an Eleatic philosopher would challenge Descartes that he defined motion in terms of motion and that he failed to give an explanation as to why motion was possible in his infinitely divisible plenum. Actually, Descartes realized the importance of foundations that offer a causal explanation for the possibility of motion

and duration in the world, as well as, for the continuous existence of created things (Franco, 2001). That was necessary
if the mechanism were to be a viable science. He understood
that an indubitable science should include motion not only
as a physical phenomenon but also as a true feature of reality. But he further understood that it was impossible to find
a causal explanation of motion in a plenum. The reason was
that an active cause is required that must be in direct contact with the body in motion and that cause should not be in
motion itself, otherwise he was defining motion in terms of
itself.

Almost 2,000 years after Zeno, Descartes understood that
it was difficult to find a primitive explanation of motion
that would not lead to a circular definition. As a result, his
mechanism was impossible. The mechanism could survive,
however, after his introduction of the *concept of continuous
recreation of the world by an immutable God.* In *Meditations
III,* Descartes argues (ibid.):

1. "I existed a little while ago"

2. "I exist now"

3. "A lifespan can be divided into countless parts, each completely independent of the others"

4. "I experience no power" [of creating myself]

Therefore, "There is some other cause which as it were creates me afresh at this moment – that is which preserves me"

The argument that Descartes presented in the *Meditations*
rests on the assumption that instantaneous existence does
not imply continuing existence unless there is a mediating
active cause. Although the argument was constructed using
the notion of *now* and of *moments of time* that are independent of each other, it nevertheless illustrates the type of conclusion one could reach when such notions are used. Those
who have attacked Descartes' argument for using indepen-

dent time divisions should not forget that in modern physics, too, such notions are fundamental. Specifically, the notion of now is necessary when taking measurements of dynamical quantities in experiments, such as speed, acceleration or force. Therefore, any attack on the soundness of the premises in Descartes' argument based on the use of undefined quantities such as "while ago", "now" and "lifespan" should not be made with the purpose of defending the soundness of the foundations of modern physics and especially those of mechanics because the same problems are present. In other words, if Descartes' argument is considered problematic because of the use of some specific temporal quantities, so are the foundations of modern physics because they are using the same quantities, whether these are defined arbitrarily or not. Of course, his conclusion could be wrong, but this is a different issue.

It is interesting at this point to examine the relation of Descartes' argument to Zeno's paradoxes, a link that has not been established so far in the literature of the history and philosophy of science. Descartes' argument in support of continuous recreation can be reformulated in the case of an arrow in flight as follows:

5. "The arrow was in motion a little while ago"

6. "The arrow is in motion now"

7. "The duration of a path can be divided into countless parts, each completely independent of the others"

8. "The arrow has no power" [of creating itself]

Zeno: Therefore, "The arrow is not moving".

Descartes: Therefore, "There is some other cause which as it were creates the arrow afresh at this moment – that is which preserves the arrow".

The link to Zeno's paradox of the arrow is intriguing. Not only Descartes' argument encompasses the original paradox of Zeno but also it attempts to resolve it by concluding that since motion is empirically observed, there must be continuous recreation of matter in discrete spacetime intervals or points. When one realizes the connection of Descartes' argument to Zeno's paradox and the solution he offers, then one could think of ways to incorporate such concepts in a theory about space and time that makes different metaphysical commitments than those made by mainstream theories. This is because one solution to Zeno's paradoxes about the impossibility of motion in plenum or vacuum is the recreation of matter in discrete spacetime, as will be discuss in more detail in Chapter Seven.

We will also discuss later in the book in more detail how Nicholas Malebranche, who was another French Cartesian philosopher, carried Descartes' doctrine of continuous recreation one step further by attributing all real causes to God. The doctrine set forward by Descartes could offer a path to the understanding of the nature of our physical reality, but its essence was undermined when Newtonian mechanics came along. It is disturbing that nothing about the Cartesian doctrine of continuous recreation is taught in schools, colleges or universities, except in a few courses offered by philosophy departments. On the contrary, concepts like motion, time, space and matter are considered self-evident and their meaning and essence is taken for granted. Yet, there are so many unresolved issues surrounding such metaphysical concepts. More importantly, as Zeno's paradoxes indicate, there is a conflict between logic and common sense approaches when using such notions. We are faced with the dilemma of either accepting logic and its axioms or rejecting it and submitting to either ad hoc intuitionism or magical accounts of the world, like particles that move in voids and forces that act instantaneously over long distances. Yet, many generations of scientists starting with Isaac Newton concentrated on models dealing at the phenomenal level of the world regardless of the fact that the fundamental notions used to construct them led to logical impossibilities.

This approach to scientific progress was useful in many respects, such as in facilitating the industrial revolution, but lacked deep explanatory content and, more importantly, contributed highly to the cultivation of a system of human values based on a specific ideology about the nature of life and the cosmos. One of the aims of this book is to contrast this ideology with the alternative that this world cannot function without intelligent interaction. In the next chapter, we will discuss some inconsistencies in the foundation of mechanics and how Newton tried to deal with them. What is more important to realize is not that these inconsistencies do exist and that they were known early in the game but rather the disturbing fact that the educational establishment has avoided to deal with them systematically.

Dynamism and the living force

Dynamism was an alternative theory developed by Gottfried Wilhelm Leibniz (1646 – 1716) with the purpose of rivaling Cartesian and Newtonian notions of force, matter, time and space. In contrast to the Cartesian view that matter is passive and has extension in space, Leibniz proposed that all there is are pointlike forces that act in such a way as to produce the phenomena in the world. Matter, space and time are the result of such pointlike interactions, which he called monads.

Let us briefly attempt to clarify the difference between the mechanistic philosophy of Descartes and Newton, called mechanics, and the rivaling dynamism of Leibniz. Mechanics postulates inert, passive matter that has extension in space and reacts with other matter by collisions and gravitation to produce observable phenomena. Newton's laws of motion and gravitation govern all material interactions. Action-at-a-distance occurs in the case of gravitational forces and between atoms of matter but mechanics actually frames no hypothesis about the nature of the mechanism involved.

Leibniz on the other hand, was against this dualistic concept of matter and force and also against the concept that force somehow arises from interactions of inert matter. Also, Leibniz argued that for motion to be real in a metaphysical sense, it must be grounded on something that is not a mere relation, something absolute and unobservable that serves as its cause (Roberts, 2003). But his most important argument was against the Newtonian views of absolute space, the arena where all the matter-force interactions take place. He argued that Newton's substantival spacetime, i.e., the notion that space and time exist independently of material things and their spatiotemporal relations, is not a well-founded phenomenon. Leibniz confronted Newtonian substantivalists with his relationism, based on which space is defined as the set of all (possible) relations among material things and that the only well-defined quantities of motion are relative ones, like relative position and relative velocity (Sklar, 1997). That contrasted Newton's notion that the only well-defined quantities of motion are the absolute ones, like absolute position, velocity and acceleration. Leibniz challenged the Newtonian notion of an absolute and unobservable space, which serves as an arena where all matter interactions take place through forces that are generated by contact, pressure, gravitation, etc. His argument was that such a concept is not well – founded because it fails empirical verification. But in another sense, Leibniz argued that such notions lead to a magical world.

In order to justify motion and interactions, Leibniz tried to identify and define an innate property of matter that would be a measurable cause of all interactions. He thus defined the living force, or *vis viva,* as the quantity mv^2, or mass times velocity squared, which he argued is conserved. But it seems that his choices were limited. The only kinematic quantity other than position that is relative and conforms to a relationist account of the world is velocity. Specifically, velocity is a frame dependent quantity. For a train passenger moving at 100 Km/hour and stationary in the train frame, his measure of vis viva is zero in his rest frame. However, an observer on the ground gets a non-zero measure based on the relative velocity of the train. Can a relative measure serve the purpose of an active force? The answer is no, unless there is magic involved in this case, too.

Leibniz tried to offer an alternative to the world of unobservable absolute space and Newtonian forces but ran into another level of problems. The main issue was that a living force cannot have many measures and thus there must be at least one measure that is metaphysically real. But how could that be? In a Leibnizian world there is no preferred reference frame to define such real value.

The efforts of Leibniz to define a quantity that is the cause of motion that does not depend on absolute space and time failed. Einstein together with Ernest Mach understood the importance of Leibniz's work and tried to revive it and account for it in their theories. In Chapter Four, we will discuss some of the inconsistencies in Albert Einstein's most famous theory, special relativity.

CHAPTER THREE: THE INCONSISTENCIES OF MECHANICS

In the previous chapter, we discussed how Descartes' mechanism failed to provide a solution to the problem of motion of celestial bodies and gravitational free fall within a quantitative and consistent framework, a task that was later accomplished by Newton. The failure was partly due to the fact that Descartes placed heavy emphasis on metaphysics rather than on quantitative models of physical reality. Descartes understood that he needed to justify motion not only as a physical phenomenon but also as a primitive notion if his science was to have a solid metaphysical foundation. Isaac Newton did not share Descartes' views about the importance of metaphysical foundations. He was more of an empiricist rather than a rationalist although it appears he understood both sides well. It is not Newton's fault that his theories were misunderstood for such a long time and are even misunderstood nowadays on a large scale. Although Newton made it clear what his science was all about, the educational establishment assigned a different essence to it. Thus, although Newton talked explicitly about mathematical concepts and models, his theories were taught as if they were laws of physical reality that described how the universe actually operates, or in other words, they were taught as laws of nature for hundreds of years. Even nowadays, high school and college students get the impression from their books and teachers that forces are real entities, that there is a phenomenon of mutual attraction of masses and even that something called gravitational mass exists. However, in Newton's Principia, where he defines the quantities used in his laws of motion and gravitation, the following statement can be found (Newton, 1952):

> I likewise call attractions and impulses, in the same sense, accelerative, and motive; and use the words attraction, impulse or propensity of any sort towards a centre, promiscuously, and indifferently, one for another; **considering those forces not physically, but mathematically:** wherefore, the reader is not to imagine, that by those words, I anywhere take upon me to define the kind, or the manner of any action, the causes or the physical reason thereof, or that I attribute forces, in a true and physical sense, to certain centres (which are only mathematical points); when at any time I happen to speak of centres as attracting, or as endued with attractive powers. (Emphasis added)

In the above statement, Newton makes it clear that his law of gravitation is a mathematical model rather than an explanatory hypothesis about how gravity works. But how many of us were made aware of this statement of Newton when we studied mechanics? Even more disconcerting is the fact that as early as in 1895, Seeliger argued that Newtonian theories implied an inconsistent universe (Vickers, 2009). Philosophers of science even nowadays cannot pinpoint the reason as to why things have turned out the way they did. They cannot figure out the reasons why the inconsistencies remained hidden for so long. Vickers (ibid.) believes that the inconsistencies remained hidden because the right question(s) were not asked. Although that may be true, I believe that it is not the deeper reason but rather it was scientific obscurantism that prevented the facts or the full details from becoming known. Is Vickers in essence trying to obscure that fact of obscurantism in science by attributing to it the wrong motive/cause? The main drive behind obscurantism in general is to preserve an established institution of knowledge from collapsing and the possible impact from that on educational but also on social and economic levels. But isn't there something similar that was going on in the middle ages with the use of scholastic philosophy?

In this chapter we will review some of the inconsistencies of Newtonian theories on a cosmological level but also at the more "local" planetary scale. The latter has been avoided persistently in physics literature because whereas inconsistencies at the cosmological scale can be dealt with the introduction of suitable modifications to a theory in the form of auxiliary hypotheses, any inconsistencies at the empirical level could be a strong indication that the theory is not a valid model of the observed phenomena, even if it has a limited domain of application.

But there is something more important than the inconsistencies and attempts to obscure them. After taking introductory physics courses in high school or college level mechanics courses, most students are left with the impression that there are such things as particles, attractive forces, gravitational and inertial mass, etc. Well, someone can argue, "What is wrong with that? When they move on and study physics in depth they will learn that Newtonian mechanics is just an approximation and it has been replaced by more advanced theories." However, this cannot be an acceptable argument to any rational person. The fact is that children grow up learning concepts that they believe relate to how the world works and that influences heavily the parallel development of a system of values that is based on such supposedly indubitable concepts they must study hard to learn. Just a tiny percentage of them, when they go on to college and elect to study physics or philosophy of science may discover the ambiguity that is inherent in the concepts they have learned in high school and then study more advanced and alternative theories, like general relativity, for example. Even then, it may be too late for them to abandon the cosmology they have already learned. For the rest, the system of the world that was already etched in their minds during childhood, based on a material world that operates autonomously through the application of forces of all kinds that are generated via interactions of inert matter, is hard to abolish. Few will manage to change their view, often to their own detriment because if they adopt a different view about the world it will not be compatible with that of the majority. This is also a fundamental reason behind the argument that students in schools should be exposed to alternatives to the theory that life started by chance. The alternatives may contribute towards the establishment of a balance of various systems of beliefs and values that currently may be the source of conflict of societies around the world.

As we shall see in the remainder of the chapter, a significant part of physics taught for the last 300 years or more is based on a host of assumptions and, although it served many useful purposes, it has little relation to physical reality and even turns out to be inconsistent at several levels. There is nothing wrong with this in principle, especially from a utilitarian point of view, and such models are indeed useful as long as people are made aware of their nature, scope and limitations and they are not elevated to a metaphysical level in such a way as to affect the development of one-sided human values while alternative views are excluded.

Newton's laws of mechanics

Isaac Newton first published his laws of motion in *Philosophiae Naturalis Principia Mathematica* (Mathematical Principles of Natural Philosophy), in 1686. *The Principia,* as it came to be known, was revised by Newton in 1713 and 1726.

Before the laws of motion were presented in *Principia,* a total of eight definitions were stated (Newton, 1952). Four of the definitions were directly related to the laws of motion, while the remaining four were used later in the treatise in connection with the derivation of the law of universal gravitation. The first definition was about the quantity of matter, the second about the quantity of motion, the third about the vis insita, or innate force of matter, and the fourth about the impressed force.

Newton's laws of motion are often stated as follows (Greenwood, 1965):

Lex I: *Every body perseveres in its state of rest, or of uniform motion in a right line, unless it is compelled to change that state by forces impressed thereon.*

Lex. II: *The alteration of motion is ever proportional to the motive force impressed; and is made in the direction of the right line in which that force is impressed.*

Lex III: *To every action there is always opposed an equal reaction; or the mutual actions of two bodies upon each other are always equal, and directed to contrary parts.*

Using modern terminology, the laws can be stated as follows (ibid.):

First Law: Every body continues in its state of rest, or of uniform motion in a straight line, unless compelled to change that state by forces acting upon it.

Second Law: The time rate of change of linear momentum of a body is proportional to the force acting upon it and occurs in the direction in which the force acts.

Third Law: To every action there is an equal and opposite reaction and thus, the mutual forces of two bodies acting upon each other are equal in magnitude and opposite in direction.

It is interesting how the modern expression of the second law of motion is derived from the original expressions of the laws and definitions of Newton. According to the definition II of Newton:

> "The quantity of motion is the measure of the same, arising from the velocity and quantity of matter conjunctly. The motion of the whole is the sum of the motions of all

the parts; and therefore in a body double in quantity, with equal velocity, the motion is double; with twice the velocity, it is quadruple."

It is clear from the above definition that Newton defined as "motion" the product of mass, and velocity, a vector that was later called momentum. In mathematical notation

$$\mathbf{p} = m\mathbf{v} \qquad (3\text{-}1)$$

where $\mathbf{p}$ is the momentum vector, m is the mass and $\mathbf{v}$ is the velocity vector. (Vectors are indicated in boldface letters).

According to the second law, the alteration of motion is proportional to the motive force impressed, or

$$\mathbf{p}_f - \mathbf{p}_i = k\mathbf{F}_m \qquad (3\text{-}2)$$

where $\mathbf{p}_f$ is the final momentum, $\mathbf{p}_i$ is the initial momentum, F_m is the motive force impressed and k is a proportionality constant. According to definition IV of Newton:

"An impressed force is an action exerted upon a body, in order to change its state, either of rest, or of moving uniformly forward in a right line. This force consists in the action only; and remains no longer in the body when the action is over. For a body maintains every new state it acquires, by its vis inertia only. Impressed forces are of different origins as from percussion, from pressure, from centripetal force."

In the above statement Newton defines an "impressed force" but it is important to notice that in the second law (Lex II) he makes use of the term "motive force impressed". What is the difference between an "impressed force" and a "motive force impressed"? A few philosophers of science have noticed this difference and some of them believe Newton expressed his second law wrongly while others argue that there is no difference between the two expressions and both lead to a tautological form of the second law, as follows:

A particle acted upon by a force will move in such a way so that the force vector will equal the time rate of change of the linear momentum vector.

Thus, if a force is applied on a particle, then its momentum will change and if the momentum changes, a force will apply to the particle. Obviously, this is a tautology. In the introduction of the graduate text "Methods of Analytical Dynamics", Donald T. Greenwood (1965) writes:

> "The concept of force as a fundamental quantity in the study of mechanics has been criticized by various scientists and philosophers of science from shortly after Newton's enunciation of the laws of motion until the present time. Briefly, the idea of a force, and a field force in particular, was considered to be an intellectual construction, which has no real existence. It is merely another name for the product of mass and acceleration, which occurs in the mathematics of solving a problem. Furthermore, the idea of force as a cause of motion should be discarded since the assumed cause and effect relationship cannot be proved.

> We shall adopt the viewpoint that the existence of contact forces can be detected and measured by springs or by other means of measuring elastic deformation. Field forces will be calculated from observed accelerations using Newton's laws of motion, or else from rules governing the force that have been established from such observations. For the most part, we shall avoid questions of cause and effect. For example, suppose a stone is tied to the end of a string and is whirled in a circular path. Does the tension

in the string cause the stone to follow the circular path, or does the motion of the circular path cause the tension on the spring? Conceivably, either viewpoint could be taken. But generally it is preferable to note that forces and accelerations occur simultaneously and neither is specifically cause and effect.

We have seen that the definitions of the fundamental units in mechanics depend to some extent upon Newton's laws of motion and these are the laws to be demonstrated. Thus, the logic has certain circularity…"

On the other hand, those who believe Newton made a mistake argue that he should have stated his second law in terms of acceleration, as follows:

The acceleration of a body is proportional to the force acting upon it and occurs in the direction in which the force acts, or in symbolic form:

$$F = ma \qquad (3\text{-}3)$$

Equation 3-3 is taught at high school level and is a special case of a more general second law also attributed to Newton. Some of those who argue about the tautological nature of the second law also reject the above form of the law as original, which, by the way, was first stated by Euler long after Newton. For others, the second law of Newton is important as it was originally stated but expressed on modern terminology as follows:

The change of linear momentum of a body is proportional to the force acting upon it and occurs in the direction in which the force acts.

This definition of the second law may be viewed as a tautology because force is equated to the change of momentum. In other words, a change in linear momentum occurs when a force acts, as follows:

$$\mathbf{p}_f - \mathbf{p}_i = k\mathbf{F}_m \qquad (3\text{-}4)$$

or, after rearranging terms

$$\mathbf{p}_f = k\mathbf{F}_m + \mathbf{p}_i \qquad (3\text{-}5)$$

and since k is just a proportionality constant, Fm must be also a momentum vector in equation 3-5 (remember we can only add apples to apples). Thus, momentum causes a change in momentum, which a tautology.

At this point and without making further assumptions it is hard to see how the celebrated fundamental law of mechanics which states that

$$\mathbf{F} = d\mathbf{p}/dt \qquad (3\text{-}6)$$

is deduced at all from Newton's Principia. However, physicists have taken the derivation for granted for more than 300 years, without questioning it. I believe this is another case of scientific obscurantism. Specifically, if we assume that equation (3-6) is the correct second law, what kind of information is missing to get to that form from the original law stated by Newton? Is this information provided in Principia by its author, Newton? These are just two of the questions one could ask. Obscurantism is the effect of an act to avoid discussing the foundational issues involved in the transition from the statements of Newton to the final form of his second law.

One answer is that we can get to the well-known and general form of second law of equation 3-6 if we remember that in Lex II Newton makes use of the expression "motive force impressed" and not just "force impressed". This is a peculiar detail that few have ever paid attention to. The only way that it appears possible to arrive at the modern form of the second law and rescue it without charging Newton with fault is to assume that "motive force impressed" and "force impressed" are two different quantities. Specifically, that "motive force impressed" is force times the interval of time in which the force is impressed and, thus, the second law becomes (Barbour, 2001):

$$\mathbf{p}_f - \mathbf{p}_i = k\Delta t \mathbf{F}_i \qquad (3\text{-}7)$$

where F_i is the impressed force. Suddenly, everything appears to fall into place because the above equation can be written as

$$\mathbf{F}_i = (\mathbf{p}_f - \mathbf{p}_i)/(k\Delta t) \qquad (3\text{-}8)$$

If we take the limit of the right hand side of the above equation as Δt approaches zero and then choose a system of units in which $k = 1$, then, from the above equation we arrive at the familiar Newton's second law:

$$\mathbf{F} = d\mathbf{p}/dt \qquad (3\text{-}9)$$

What was essentially done in an effort to "save the second law" was to equate the "motive force" to a quantity defined in physics as impulse, which is equal to the product of force times time. Thus, if we take this route, the agent that changes momentum from an initial to a final state is a physical quantity called impulse and it is equal to the product of force times time.

One can see this, for example, in the case of a solid steel block resting on a surface with friction. Application of a force on the block in a direction parallel to the surface may not suffice to move it. This kind of force is impressed but not motive. It becomes motive when its magnitude is large enough to overcome friction and cause motion. Thus, a motive force is one that is causing motion.

Newton's second law: physics or metaphysics?

In the previous section, we showed how one could derive the Newton's second law from his original statement of it. The extra step we have taken was to introduce the concept of time in the derivation of the second law and specifically in equation 3-7. The importance and implications of this step should not be underestimated. If Newton's second law is to be correct, a motive force must be the product of force times the interval of time in which it is impressed on a body. This is a highly problematic concept and although it salvages Newton's second law, it raises important philosophical questions. For example, one such question is: isn't time just a human invention or it is a feature of physical reality? If time does not exist as a physical dimension and it is just a human invention, then what does the product of force times time actually mean in the laws of motion?

Remember that we are talking about laws of motion, at least that is what we were told when we studied them. We are not talking just about a useful device to estimate force or acceleration of a body in motion and determine its trajectory in space. We further realize that force can only be measured by measuring its effect, like the tension on a spring or the change in the velocity of a body. To say that there is a force that causes a certain effect is equal to making a metaphysical commitment to a true cause. In other words, in the original Newton's second law, equation 3-7, on the left-hand side we have the change in momentum, which is an empirically measured quantity. However, on the right hand side we have force times time and if time and force are both metaphysical commitments then we set an empirical and well-defined quantity equal to a metaphysical quantity, in the following sense:

Empirical quantity = metaphysical quantity

What does it actually mean when one equates an empirical quantity to a metaphysical quantity? One possible answer is that this kind of statement is just a mathematical model, an instrument to make predictions of some sort, and not a law of nature. This is what many scientists will argue but this is not what people have been told for hundreds of years and are still told in schools, which is the main point made here. Another possibility is that the empirical world, or the world of phenomena, is related to a real world in a metaphysical sense, where the true features of reality have real existence and ontology and are the true cause of everything. This Platonic view has been rejected long ago by empirical science as not necessary for the study of the phenomena of nature. However, those same scientists claim that Newton's second law is an empirical principle while some others claim that it is a law of nature. How can a law that equates an empirical quantity to a metaphysical notion be empirical? Unless, one believes that time and force are also empirical quantities. Well, force can be measured indirectly using spring scales. But that can hardly justify using the concept of force in empirical science since what can be measured is only the effects attributed to forces and such effects can actually be the result of some other type of interaction that is not also observable.

Furthermore, time is measured by clocks that either involve harmonic motion, such as a pendulum, or count events, such as atom transitions. Time is thus measured in terms of motion or event count. That again does not prove that such a dimension exists in a fundamental sense. Newton attempted to circumvent this problem by defining true time as something absolute that flows independently of the other three spatial dimensions (Newton, 1952):

> Absolute, true, and mathematical time, of itself, and from its own nature flows equably without regard to anything external, and by another name is called duration: relative,

> apparent, and common time, is some sensible and external (whether accurate or unequable) measure of duration by the means of motion, which is commonly used instead of true time; such as an hour, a day, a month, a year.

The term "true time" alludes to something metaphysical, i.e., it signifies a true feature of reality and it is due to this fact that empirical scientists have long since surrendered to the simpler notion that time is "what clocks measure". However, the proposition that "time is what clocks measure" is analytic, and involves circularity. This by no means affects the usefulness of physical theories but it raises questions regarding their meaning, their relevance to physical reality and about the way they are presented in the educational curriculum.

Some may argue that the issues just discussed have no value for physics because they deal with metaphysical questions and their answers are not required to solve practical problems. Others may assert that any answers to these questions, if they exist at all and can be found, may play a decisive role in understanding our physical reality. I will not try to argue in favor of either side at this point. In this chapter, I offered a brief introduction to the problems one is faced with by just considering the most fundamental law everyone learns in high school, Newton's second law of motion. Even the derivation of the law from the original statements by its inventor is problematic unless some fundamental assumptions are made. Things are much more complicated than most scientists and teachers would like people to believe. Nevertheless, I will not try to discuss these specific philosophical issues in more detail in this section but move ahead and present some inconsistencies of Newtonian mechanics that raise several other important questions.

Inconsistencies of Newtonian mechanics

Before presenting a few inconsistencies of Newtonian mechanics, it is worthwhile to review Newton's law of universal gravitation.

One of Newton's major accomplishments was the formulation of his law of universal gravitation. This famous law is expressed — in its simplest form — as follows:

$$F(r) = -\frac{G\,m_1\,m_2}{r^2} \qquad (3-10)$$

Newton's gravity law states that two particles of mass m_1 and m_2 separated by a distance r attract each other with a force that is proportional to the product of the two masses and inversely proportional to the square of the distance, as expressed by equation 3-10. The proportionality constant G is called the *constant of universal gravitation.* The minus sign in 3-10 indicates that $F(r)$ is an attractive force. The value of the constant G is approximately equal to 6.671 x 10-8 m2/Kg sec2 and assumed to be independent of all masses present or their location and orientation in space.

The time-reversibility inconsistency

Recall that Newton's second law reduces to $\mathbf{F} = m\mathbf{a}$, for constant mass m, when a force $\mathbf{F}$ is applied to result in acceleration $\mathbf{a}$.

The acceleration a is given by:

$$\mathbf{a} = \mathbf{F}/m \qquad (3-11)$$

We next set m = 1 in equation (3-11) for simplicity and in appropriate units. Since $\mathbf{a}$ = d$\mathbf{v}$/dt, the above equation is equivalent to:

$$d\mathbf{v} = \mathbf{F}dt \qquad (3\text{-}12)$$

If we assume for simplicity a zero initial velocity, after applying integration using equation 3-12, we get:

$$\mathbf{v} = \mathbf{F}t \qquad (3\text{-}13)$$

Since $\mathbf{v}$ = d$\mathbf{s}$/dt, the above equation yields:

$$d\mathbf{s} = \mathbf{F}tdt \qquad (3\text{-}14)$$

Assuming that $\mathbf{F}$ is constant and initial conditions are equal to zero, we can integrate both sides of equation (3-14) twice to obtain the following equation for position as a function of time t in the case of linear motion:

$$s = Ft^2/2 \qquad (3\text{-}15)$$

The above equation is time-reversible, in the sense that the position s is a function of t squared. Let us assume for simplicity that F = 1 in proper units. Then, the magnitude of the distance is given by:

$$s = t^2/2 \qquad (3\text{-}16)$$

We may see that s = 2 units of distance can be achieved after t = 2 or t = -2 seconds. There is nothing in Newton's second law that demands that when a force F is applied to a mass m, it will reach a distance s in positive time, meaning in the future, or in negative time, meaning in the past.

Although an attempt is made to justify the existence of an arrow of time on a universal scale based on thermodynamic considerations, such as the increase of the total entropy of the universe, this cannot explain why there is an arrow of time at the particle level. There are some naïve answers in this area. One is that movement in the past is possible but its probability is low. Such naïve claims raise more questions than they answer, like for example, why we do not have incidents of things disappearing in the past when a force applies on them, or, for example, when applying a force to a small mass repeatedly for thousands of times or even when we press on the gas pedal of a car. The claim that the probability of something like that happening is so small that it can hardly happen is unacceptable. Since the beginning of the industrial revolution we have mechanically generated and applied forces on all sorts of bodies in laboratories, and in everyday life, countless times. There is not a single instant of reported time reversibility. The argument from probability appears to be a desperate effort to salvage the autonomy of this world. The same scientists that reject the small probability argument made by Dembski (2002) in favor of intelligent design accept the low probability argument against time irreversibility in order to salvage the laws of mechanics. This is certainly a dangerous eclectic stand. If an arrow of time is imposed on the phenomena that essentially guarantees that the small probability of time reversibility, as allowed by the laws of motion, is in reality equal to 0, then it is not immediately evident how that can happen in the context of an autonomous universe. Certainly, the fact that there are irreversible processes does not imply that there is an arrow of time imposed on all phenomena. This is the reason that the time reversible nature of the equations of motion of mechanics is an inconsistency because it allows the proposition $p = \{\text{particle moves in time t}\}$ and the proposition $\sim p = \{\text{particle moves in time } -t\}$ to be derived simultaneously.

If modern science believes that there is an arrow of time compatible with an autonomous universe, then where are the equations of motion that are compatible with such a doctrine? There are none available in reality. In the presence of acceleration, calculation of the position vector will entail time reversibility. Classical mechanics laws are inherently inconsistent and problematic and so are the modern theories which are based on it, irrespectively of whether that happens at the strong or at the weak field limit. For example, general relativity must converge to equations 3-10 and 3-11 at the weak field limit, as for example at the planetary system level. Thus, any appeal to general relativity as having a wider application and domain than Newtonian classical mechanics does not resolve the time reversibility inconsistency.

Finite versus infinite universe inconsistency

Infinite universe

If the universe is infinite and contains infinite matter, then Newtonian gravitational potential would be infinite at every point in it. Albert Einstein wrote:

> According to the theory of Newton, the number of "lines of force" which come from infinity and terminate in a mass m is proportional to the mass m. If, on the average, the mass-density $\rho 0$ is constant throughout the universe, then a sphere of volume V will enclose the average mass $\rho 0V$. Thus the number of lines of force passing through the surface F of the sphere into its interior is proportional to $\rho 0R$. For unit area of the surface of the sphere the number of lines of force which enters the sphere is thus proportional to $\rho 0V/ F$ or to $\rho 0R$. Hence the intensity of the field at the surface would ultimately become infinite with increasing radius R of the sphere, which is impossible. (Einstein, 1917, p.106)

Vickers (2008) counters Einstein's claim on the basis that potential is a non-physical quantity and as a result it does not matter if it is infinite. What matters, according to Vickers, is whether the force can be determined at each point in the universe or not. Such a counter argument makes an implicit unfounded assumption that forces are physical entities and gravitational potentials are not. However, there is no basis for such assumption, as discussed earlier, because the ontology of physical reality is unknown. The choice of force or potential as the basis for evaluating the Newtonian theory for consistency is arbitrary and carries no specific relation to reality. What Vickers (ibid.) misses that is more important is that Einstein cared about the consistency of the mathematical model of Newtonian mechanics and not whether such model represents physical reality. If you consider forces directly, you get, as Vickers correctly shows, an indeterminate force on every test mass in a Newtonian universe but if you do the same analysis using gravitational potential instead, you arrive at infinite potentials.

More importantly, Vickers (ibid.) seems to underestimate the significance of the fact that in the case of Newtonian mechanics a parameter becomes infinite in the absence of relativistic conditions, namely the potential, which was apparently what bothered Einstein. In order to understand this, let us consider a test mass on which two forces of the same magnitude act in opposite directions. According to Newton's second law of motion, the test mass should maintain its state of rest or uniform linear motion. Now, if the magnitude of the forces becomes infinite, can we say that the same conclusion should hold but the net force on the test mass is indeterminate? We know from mathematics that the operation infinity minus infinity is not defined and leads to a contradiction if one assumes it has a result. The contradiction is illustrated as follows:

Let us assume that:

$$\infty - \infty = 0$$

Now we add the number one to both sides of the above equation:

$$\infty - \infty + 1 = 0 + 1$$

But

$$\infty + 1 = \infty \quad \text{and} \quad 0 + 1 = 1$$

and therefore:

$$\infty - \infty = 1$$

We have thus demonstrated with the above trivial example that indeterminacy results in a contradiction if we make the assumption that an operation that involves infinities has a given but undetermined value, because in reality it can have any value.

The problems with indeterminate forces and infinite potentials motivated some scientists to consider modifications to Newtonian gravity, including a diminishing gravitational potential as a function of distance (Milgrom, 2002). However, none of the modifications could fully explain the motion and formation of galaxies and distant stars, so scientists were forced to contemplate that the universe is finite.

Finite universe

The problem with a finite universe is that the forces of gravity would cause it to contract to a single point. Albert Einstein looked at this problem in depth and he decided at some point to consider what he called a *cosmological constant* in order to guarantee a stationary universe, although he later abandoned the idea. Recent observations have shown that the universe expands at an accelerating rate and satellite measurements support the cosmological constant hypothesis.

One major inconsistency in the finite universe model arises from the fact that modern theories predict a huge cosmological constant whereas observations result in a small value (Weinberg, 1989). Actually, the difference is in the order of 10^{100}!

Thus, we see that an infinite universe in mass and scale leads to inconsistencies in the framework of Newtonian mechanics, but also a finite universe is problematic unless one makes assumptions that are in conflict with empirical evidence by a huge factor. There are also some bizarre theories that do not require a cosmological constant but resort to peculiar notions like parallel universes and dark energy.

The point in this book is not which theory is correct but the fact that there is so much ambiguity and conflict in modern physics that results in a science that cannot be portrayed in schools and universities as anything else but a mix of hypotheses and experimental data. *On the contrary, students are driven to believe that science thrives on the accumulation of concrete knowledge that builds on rigid foundations.* Nothing can be further from the truth and those that pass this message are either naïve or liars. Nowadays, we are faced with the same problem Albert Einstein faced 100 years ago or even Newton about 300 years ago. There is nothing concrete in science and the foundations can collapse at any time, as the argument of pessimistic meta-induction claims (Laudan, 1981).

The total eclipse of the sun inconsistency

We next consider a more practical situation: the case of a total eclipse of the sun, occurring when the sun-moon-earth, a three-body system, lines up in space, as shown in figure 3-1. The following values are used in the calculation of the forces Fs and Fe on the moon arising from the gravitational attraction of the sun and earth, respectively:

Moon distance from earth, R_e = 3.84 x 108 m
Moon distance from sun, R_s = 1.49 x 1011 m
Mass of earth, M_e = 5.98 x 1024 kg
Mass of sun, M_s = 1.99 x 1030 Kg

According to Newton's law of universal gravitation, the force on the moon due to the sun is:

$$F_s = \frac{GM_m M_s}{R_s^2} \qquad (3-17)$$

Similarly, the force on the moon due to earth is:

$$F_e = \frac{GM_m M_e}{R_e^2} \qquad (3-18)$$

Dividing equations 3-17 and 3-18 yields:

$$\frac{F_s}{F_e} = \frac{R_e^2}{R_s^2} \frac{M_s}{M_e} \qquad (3-19)$$

Substituting in 3-19 the values for the mass of sun and earth and the corresponding distances to the moon yields:

$$\frac{F_s}{F_e} = \frac{(3.84 \times 10^8 \text{ m})^2}{(1.49 \times 10^{11} \text{ m})^2} \frac{1.99 \times 10^{30} \text{ Kg}}{5.98 \times 10^{24} \text{ Kg}}$$

$$= 2.21$$

or

$$F_s = 2.21 F_e$$

Therefore, the force on the moon due to sun's pull is 2.21 times the force due to earth's pull. Notice that neither the value of the mass of the moon nor the value of the gravitational constant G was required to get the final result.

At the point Pe of the total sun eclipse, as indicated on Figure 3-1, the three bodies line up and the moon has a tangential velocity Vm due to its trajectory in space. Since the net force on the moon, F_s-F_e, is in the direction of the sun center, then according to Newtonian mechanics, the moon should have some type of orbit around the sun defined by one of the four conic sections specified by the solution of the two-body problem, or a singular solution representing a collision of the two bodies. If conditions would allow it, the moon could become a planet orbiting the sun rather than the earth.

A departure of the moon from earth's orbit has not happened during a total eclipse of the sun and this is just one indication that Newton's inverse squared law, when combined with the laws of motion under certain unique and rare conditions, does not describe what actually happens. Some will attempt to justify this failure by using "creative terminology" such as "the earth-moon system falls toward the sun" or that the "moon is in earth's sphere of influence". These sound more like poetry than science. The situation described is inconsistent with Newtonian mechanics and any use of creative terminology cannot change this conclusion. If there is something else preventing the moon from orbiting the sun only like a regular planet, then it is certainly not directly deduced from Newton's laws but additional assumptions must be considered.

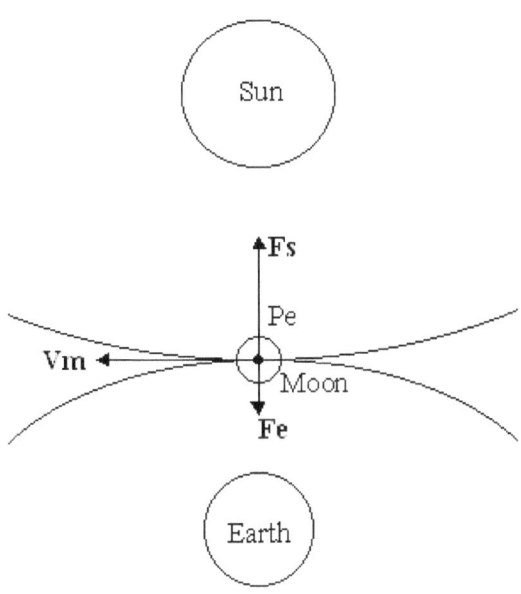

Figure 3-1. Forces on the moon during total eclipse of the sun

It appears that the moon's orbit around earth is some type of inertial motion that Newton's laws of motion cannot explain but only describe its kinematics. This is probably a reason this inconsistency has been persistently obscured. Its existence challenges directly this world's autonomy and points to an alternative system of laws of physics according to which motion in arbitrary closed paths is inertial. This was actually accomplished in a way by Einstein's general relativity, but in a complicated way that completely removes causality in the case of gravitation. This is because Einstein's theory applies in four-dimensional spacetime only. In this book, however, I will present a system of laws of motion more general than Newton's that can account for inertial motion on closed arbitrary three-dimensional paths. These laws of motion make significantly different metaphysical commitments than Newton's laws and may indicate that there are alternatives that may have been obscured.

More importantly, those who insist there is no inconsistency in the total eclipse situation and claim so by inventing terminology and making all sorts of auxiliary hypotheses are not aware of, or maybe are even trying to obscure the fact that Newton knew that his mechanics only modeled the orbits of celestial bodies but could not explain them. Newton (1952) wrote:

> The six primary planets are revolved about the sun in circles concentric with the sun, and with motions directed towards the same parts, and almost in the same plane. Ten moons are revolved about the earth, Jupiter and Saturn, in circles concentric with them, with the same direction of motion, and nearly in the planes of the orbits of those planets; **but it is not to be conceived that mere mechanical causes could give birth to so many regular motions....** This most beautiful system of the sun, planets, and comets, could only proceed from the counsel and dominion of an intelligent and powerful Being. (Emphasis added).

Basically in the above statement Newton argues that mere mechanical causes could not give birth to the planet orbits but it can only be the result of the action of an "an intelligent and powerful Being". This is because Newton understood that his laws were not enough to "save the phenomena" but only to describe them. And this is the key here, i.e. the difference between laws that describe phenomena and laws that explain phenomena.

Without going into some elaborate details, the inconsistency just discussed arises if both the laws of motion and the law of universal gravitation are to apply consistently under all circumstances, as in the total eclipse situation. In order to settle things, Newton made the following hypothesis and associated proposition in his Principia (ibid.):

HYPOTHESIS I.

That the center of the system of the world is immovable.

This is acknowledged by all, while some contend that the earth, others that the sun is fixed in that center. Let us see what may from hence follow.

PROPOSITION XI. THEOREM XI.

That the common center of gravity of the earth, the sun, and all the planets, is immovable.

For (by Cor. 4 of the laws) that center either is at rest, or moves uniformly forward in a right line; but if that center moved, the center of the world would move also, against the Hypothesis (I).

We see that the proof of theorem XI is based on hypothesis I. It is clear that Newton's law of universal gravitation rests on a hypothesis concerning the existence of a common center of gravity, which is immovable, which in turn is justified by the hypothesis about an immovable center of the system of the world. It is kind of peculiar that the use of such a hypothesis is in a contradiction with Newton's closing remarks regarding the cause of gravity and the use of hypotheses, in general (ibid.):

But hitherto I have not been able to discover the cause of those properties of gravity from phenomena, and I frame no hypotheses; for whatever is not deduced from the phenomena is to be called an hypothesis; **and hypotheses, whether metaphysical or physical, whether of occult qualities or mechanical, have no place in experimen-**

tal philosophy. In this philosophy particular propositions are inferred from the phenomena, and afterwards rendered general by induction. Thus it was that the impenetrability, the mobility, and the impulsive force of bodies, and the laws of motion and of gravitation, were discovered. And to us it is enough that gravity does really exist, and act according to the laws which we have explained, and abundantly serves to account for all the motions of the celestial bodies, and of our sea. (Emphasis added).

The Newtonian model was useful in making fairly accurate predictions in the absence of relativistic effects but it does not provide a deep understanding of how physical reality operates. Models such as Newton's have been useful to progress since the time of Archimedes. Problems arise, however, when a real identity is attributed to the variables used in the model, such as to a gravity force and to the gravitational mass. The real existence of both of these quantities is not justified but their invention and their use to develop models of physical reality were an achievement. Instead of focusing on the essence of that brilliant methodology in the curriculum of educational institutions, the model itself has been taught as a law of physical reality in a totally misrepresented and incorrect way and students have been given the impression that force and gravitational mass truly exist and have an identity in a physical sense. I am certain that if Newton were to attend a classical mechanics lecture dealing with the subject of his laws he would be really amused by the interpretations given to his work. At the time of Newton and Leibniz, philosophy and science were at a high and elevated balance and that allowed progress to be made. In a certain sense that maybe is not fully understandable nowadays. As a matter of fact, Leibniz was a fierce critic of Newton and made an effort to come up with a system of the world that did not require such things as Newtonian forces, absolute space and time, but was based on a purely relational account. At the same time, Newton heavily criticized Leibniz's approach and even presented experimental results that according to him were evidence of the existence of an absolute space (Sklar, 1997). It suffices to say, that even nowadays, the debate of the nature of space and time started by Newton and Leibniz has not been settled. In the second part of the book we will discuss a proposal for a final settlement. What is important to understand is that the educational establishment already decided many years ago which side to take without having had any evidence at all. The issues raised by relativity, as we shall also see in the next chapter, were dismissed for the last 100 years and instead a distorted version of Newtonian mechanics is being taught that was intended to provide certainty about the nature of the world, whereas such certainty does not exist. As a matter of fact, in the second part of this book I will show that Newton's laws are a subset of a more general set of laws of motion derived from a single principle. The metaphysical implications from this discovery are profound and directly challenge the autonomy of this world by requiring continuous and intelligent interaction so that the perceived causality is not violated at the level of the phenomena. In other words, there is an alternative model of this world that is compatible with a non-autonomous reality. If reality is non-autonomous but requires intervention for its existence then notions of an emergence of complicated physical forms by chance are problematic, to say the least.

I would like to make it clear again that I do not take sides although I am sympathetic towards the intelligent interaction hypothesis. What is at stake here is the cultivation of a one-sided system of values based on a specific cosmology that precludes non-autonomy of the world but ascribes all phenomena to some physical laws without offering any clues how such laws emerged and persist in the universe. Instead, modern science resorts to auxiliary hypotheses when its models break down, such as, for example, the existence of some field that attributes mass to all particles in the universe, and allocating tremendous resources to its pursuit when the unemployment in some of the countries that fund this research is above 20%. If one looks closer at what is happening in modern science it is evident that a proof of the autonomy of the world must be chased at any cost while at the same time alternative views that also support different system of values must be suppressed as being non-scientific. This book is a first and small step towards the establishment of a scientific foundation for theories that support intelligent interaction, intervention and design. The journey of corroborating or falsifying such theories may be long but long journeys are always initiated by small steps.

CHAPTER FOUR:
INCONSISTENCIES OF SPECIAL RELATIVITY

Relativity in physics is nowadays synonymous with special relativity, the theory developed by Albert Einstein and presented in his 1905 paper "On the Electrodynamics of Rigid Bodies" (Einstein, 1905). There are other scientific theories that fit the label "relativity theory", including general relativity, but in this chapter we will deal only with some specific aspects of special relativity since this is one of the most widely accepted theories by the scientific community for studying phenomena where gravity does not play an important role and the speeds get close to the speed of light.

In a nutshell, special relativity is a theory that applies a specific form of a principle of relativity to a group a special reference frames, called global inertial reference frames. In place of the common sense view of space and time being absolute and universal, respectively, special relativity proposes a four-dimensional spacetime where different observers measure different times and locations for the same event, depending on their state of motion. As it turns out in special relativity, simultaneity is relative to a reference frame. Special relativity converges to Galilean relativity, which is the basis of Newtonian mechanics, when speeds are low compared to the speed of light. These concepts may not be intuitive but this is not the main issue here. The issue of importance is whether this widely accepted theory is consistent, or not.

A central notion in special relativity is the spacetime interval between the origin event (0,0,0,0) and an event (x,y,z,t), defined as follows:

$$s^2 = x^2 + y^2 + z^2 - (ct)^2 \quad (4\text{-}1)$$

The spacetime interval is invariant for all observers in inertial reference frames irrespectively of their speed. Note that in three-dimensional space the term that involves time in equation (4-1) is not present. In special relativity, time is an imaginary coordinate and part of a four-dimensional spacetime, also known as *Minkowski spacetime.* Equation (4-1) looks like an ordinary mathematical object but when used in a theory of physics, it establishes a link between the empirical world of rulers and clocks that allow making local measurements in inertial reference frames and the metaphysics of a four-dimensional spacetime. This equation reflects a metaphysical commitment made by a more accurate model of physical reality in its domain of application.

These are the two fundamental principles of special relativity that when taken together distinguish it from other relativity theories:

First principle *(principle of relativity):* The laws of physics are the same in any inertial reference frame, regardless of position or velocity.

Second principle *(constancy of the speed of light):* The speed of light is the same in any inertial reference frame, independent of the speed of its source.

The two fundamental principles, together with additional principles dealing with the geometry of space and other basic assumptions comprise the formal axiomatic foundation of special relativity, from which all its theorems and predictions are derived.

We will not go into the details of special relativity here, but only briefly discuss how the principle of relativity, combined

with the constancy of the speed of light lead to some contradictions within the domain of application of that theory
and how the theory attempts to deal with them. Generally
speaking, a theory is inconsistent when it allows deriving a
conclusion p and also its negation ~p.

One of the other two principles of Einstein's special theory of relativity is that space and time form a four-dimensional continuum, called spacetime, as defined by equation
(4-1). This principle is assumed to be true but it amounts
to a metaphysical commitment. One cannot directly prove
or falsify this principle because spacetime in that particular
multidimensional form is not directly observable. Thus, we
can immediately see that this theory poses an unobservable
entity but in conjunction with other postulates and empirical
principles, Einstein's theory of special relativity, in its modern formulation, generates many testable predictions that
have been verified in laboratories numerous times. So far,
none of the predictions made by special relativity have been
falsified in a laboratory experiment although scientists have
intensely targeted them and the theory is accepted as a valid,
high precision model in its domain of application. Yet, when
common sense notions of space and time are employed, use
of this theory leads to contradictions.

The most famous contradiction deduced from special relativity is the twin paradox. Since there is quite a controversy
surrounding this paradox and its proposed solutions, I will
present two related situations. These situations illustrate the
same problem, i.e., how the principle of relativity and the
constancy of the speed of light considered together in the
same theory lead to logical contradictions. Attempts to resolve the contradictions lead to a peculiar notion of physical
reality. Note, however, that these contradictions do not constitute proof that the theory is wrong, as will be discussed
later in this chapter. I would like to make this clear in ad-
vance.

The pole-barn contradiction

Figure 5-1 illustrates this problem leading to a contradiction. There is a pole and a barn both at rest in an inertial reference frame. The pole is longer than the barn width and cannot fit in it. I do not call this the "pole-barn paradox", because the use of the term paradox is only acceptable if one has already determined that this appears to be a paradox but actually it is not.

The pole is accelerated to a constant speed near the speed of light. According to special relativity, the pole contracts as seen from the frame of reference of the barn, and by a proper selection of the pole length, the barn width and the pole speed it can be made to fit in the barn.

However, according to the principle of relativity, there is no preferred inertial reference frame and one can consider the same pole at rest and the barn moving at the same constant speed towards it, as shown in figure 5-1. In that case the barn contracts and the pole cannot fit in it. Obviously, we have two contradictory conclusions derived from the same theory and that is an indication that the theory may be inconsistent.

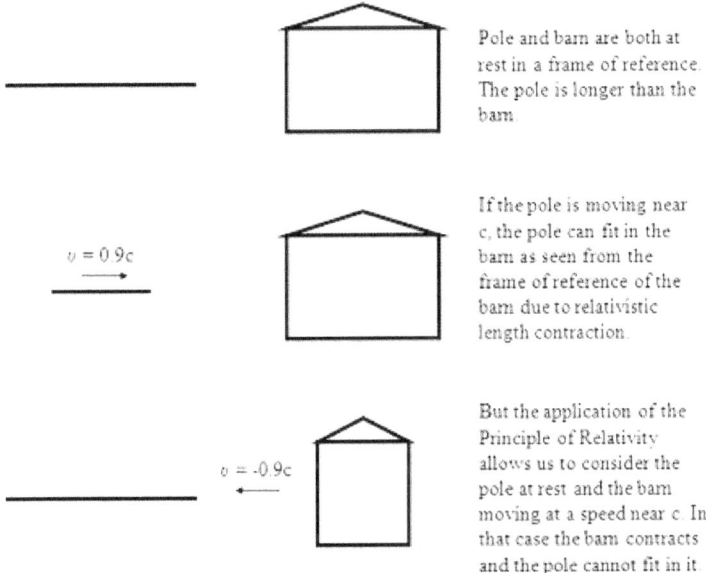

Pole and barn are both at rest in a frame of reference. The pole is longer than the barn.

$v = 0.9c$

If the pole is moving near c, the pole can fit in the barn as seen from the frame of reference of the barn due to relativistic length contraction.

$v = -0.9c$

But the application of the Principle of Relativity allows us to consider the pole at rest and the barn moving at a speed near c. In that case the barn contracts and the pole cannot fit in it.

Figure 5-1. The pole-barn inconsistency (http://hyperphysics.phy-astr.gsu.edu/hbase/relativ/polebarn.html)

The bug-rivet contradiction

Figure 5-2 illustrates this contradiction. There is a bug at the bottom of a hole and a rivet, all initially at rest in an inertial reference frame. The rivet is shorter than the depth of the hole. Then, the rivet is accelerated at a constant speed close to the speed of light towards the hole. The bug sees the rivet contracting in accordance to special relativity and does not worry about getting squashed when it reaches the hole.

However, according to the principle of relativity, one can consider the rivet as being stationary and the hole moving at the same but opposite constant speed near c. In this case, the hole (and bug) contract and the bug will get squashed. Again, we have two contradictory conclusions derived from the same theory and that may be an indication that the theory is inconsistent.

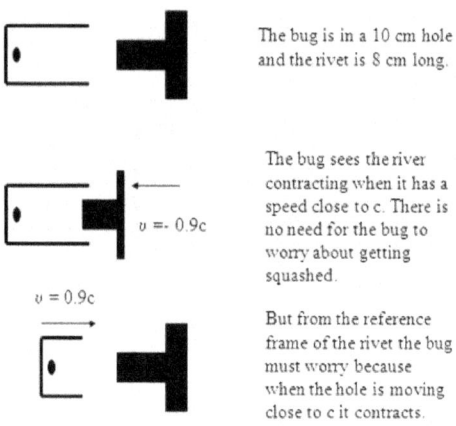

The bug is in a 10 cm hole and the rivet is 8 cm long.

The bug sees the rivet contracting when it has a speed close to c. There is no need for the bug to worry about getting squashed.

But from the reference frame of the rivet the bug must worry because when the hole is moving close to c it contracts.

Figure 5-2. The bug-rivet inconsistency (http://hyperphysics.phy-astr.gsu.edu/hbase/relativ/bugrivet.html#c1)

Some have argued that the above situations do not prove special relativity is inconsistent but that they are paradoxes that are resolved when one realizes that they describe conditions that cannot happen near the speed of light. For example, no bug can survive moving at the speed of light, that no hole can sustain contact forces that are generated at such speeds and a host of other arguments in favor of an apparent paradox. However, this is naive thinking. The objective of the two examples described above was to illustrate a foundational problem of a theory. One can consider more realistic situations that involve moving particles and detectors that send signals before getting destroyed.

Does this mean that special relativity is inconsistent and false? We cannot answer this question unless we know the true nature of our world. It may be the case that the bug both gets squashed and does not get squashed, depending on the reference frame and that both these situations are possible and exist independently as two distinct probable events in two probable worlds. Furthermore, it may actually be the case that an inconsistent theory better describes our world because it is also inconsistent. This may be a strange world, as Einstein said once in a letter to a friend. However, the inconsistencies may be avoided if we consider some bizarre physical reality.

The high price for avoiding the inconsistencies

Relativity, and in particular special relativity, changed the classical views of a world where substances with intrinsic properties, such as a body with extension, move in three-dimensional space where there is objective simultaneity of events. This change however came at a high price and most scientists are still not convinced it has to be paid. Most accept the relativity theory as an accurate mathematical model in an instrumentalist sense but not its metaphysical commitments, in a realist sense. As a result, they doubt that relativity is a true description of physical reality.

Specifically, relativity commits to a tensless theory of time and existence and to events instead of substances (Esfeld, 2009). This is a drastic departure from the metaphysical commitments of classical physics. According to relativity, simultaneity is only relative to a reference frame and is not objective. As a result, an event that it is the past of an observer can be in the future of another observer depending on their relative state of motion. Furthermore, everything that exists in the past, present or future, exists in a four-dimensional block universe.

The objects in the block universe are different from the objects of classical mechanics and of a three-dimensional universe. The objects of the relativity theory have also temporal parts, not just geometric. Their extension is not fixed, as in the case of rigid three-dimensional objects, but is dependent on the reference frame of the observer. In the relativity theory, physical objects are continuous sequences of spacetime points, also known as processes. According to this radical departure from classical mechanics, motion in a block universe can be defined as a continuous sequence of spacetime points that have similar physical content. Change is a continuous sequence of spacetime points with different physical content (ibid.).

There are those who will argue that the notion of three-dimensional objects with extension is compatible with special relativity. This view leads to inconsistencies, like the bug-rivet paradox, already discussed in this chapter. One is forced to accept only four-dimensional objects in special relativity, so that they retain their physical shape relative to all inertial reference frames. Thus, objects must have temporal parts in addition to physical extension to avoid inconsistencies. This is how the two inconsistencies discussed in this chapter can be resolved. However, if there is actually a preferred reference frame it means special relativity is not a true theory of nature. Those that argue that special relativity is an inconsistent theory because the principle of the constancy of the speed of light and the principle of relativity combined lead to contradictions, may be compelled to do so because they commit to the existence of three-dimensional objects. Actually, Albert Einstein (1916) was correct to include the constancy of the speed of light postulate because it made the theory consistent. It is another matter whether special relativity is a true theory of physical reality or if it is just an instrument for making predictions. If there is an unobservable preferred reference frame which serves as the arena in which three-dimensional objects take part in events, then, in that case, relativity is just a useful model.

The main objective of this book is to demonstrate the ambiguity that plagues contemporary science and the fact that things are not as certain as they are often presented by the educational establishment. Actually, the establishment has taken a dogmatic stand against the questioning of foundations in the same way that medieval priests treated any questioning of religious doctrines. But there are serious problems with the foundations of science and the use of naïve perceptions of space, time and matter, as the Eleatic philosophers warned us long ago. We discussed some of these early warnings in Chapter One. In this chapter we showed how the same naïve conceptions induce inconsistencies in the theory of special relativity and the high price we must pay in order to avoid them, which is called four-dimensionalism. It may thus be worthwhile to consider alternative foundations based on a completely different notion of reality and its relation to the perceived phenomena. This is what is attempted in the second part of the book. In the case of the alternative presented, the price paid is the loss of the autonomy of the world. Thus, on one side we have theories like relativity and loss of presentism and free will to gain quasi-autonomy and, on the other hand, the alternative presented in the second part of this book and the loss of autonomy to preserve presentism and some sort of free will. The two alternatives balance each other and present us with a broader spectrum of choices. However, they also commit to different ideologies that often serve us the foundations from which systems of values emanate.

CHAPTER FIVE: THE SPOOKINESS OF QUANTUM MECHANICS

The quantum theory is considered by many physicists to be one of the most successful theories of all times. The predictions that this theory makes are precise and have been confirmed to high accuracy numerous times in many laboratories around the world. It is believed that about one third of the gross national product of the USA is directly or indirectly related to advancements in quantum theory (Lyre, 2010). This chapter is a very brief introduction to some of the foundational issues of quantum theory. Due to the complexity of these issues, an effort has been made to present them in way that does not overwhelm the reader but at the same time conveys the main message of this book, i.e., that the foundations of modern science are not as rigid or as clear as some scientists want us to believe. In many cases, the line that separates a mechanistic world from a world of miracles is not entirely clear. This is especially true with quantum theory.

In quantum theory, the fundamental law is the time–dependent form of the Schrödinger equation:

$$i\hbar \frac{\partial}{\partial t} \psi(r,t) = \left[\frac{-\hbar^2}{2\mu}\nabla^2 + V(r,t)\right]\psi(r,t) \qquad (5\text{-}1)$$

where ψ is the wave function, i is the imaginary unit, $\hbar$ is the Planck constant divided by 2π, μ is the reduced mass of the particle, and V is the potential energy. This law is expressed in the form of a partial differential equation that describes the time evolution of a quantum system's wave function. How this equation originated is not fully known. Richard Feynman thought the equation was an intellectual construction and that it was impossible to derive it from anything that was known (Hey, 2009).

As with Newton's second law and its commitment to the metaphysics of force, in a similar but more intricate way, Schrödinger's equation is a mathematical object that relates empirical measurements to the metaphysics of a wave function. Although there is the standard interpretation of quantum theory in which the wave function is an intellectual construction that is useful for analyzing quantum systems, there are several other interpretations that assign a real existence to it. All of our physical theories use mathematical models this way, i.e., by linking a primitive ontology to nomological variables. In the case of quantum theory a nomological variable is the wave function and in the case of Newtonian mechanics it is the momentum. The link to nomological variables is necessary because physics goes beyond assumptions about a primitive ontology and aims at describing how reality evolves in time. In simpler words, since we do not know the true nature of our physical reality, we conceive certain mathematical laws that link some (hypothetical) features of it that we think are primitive, such as for example particles, to some other features that are not primitive, such as for example momentum or wave function. The non-primitive features are necessary for describing how physical reality evolves in time. Without the notion of momentum it would be impossible to derive Newton's second law despite of an assumed primitive ontology. However, the choice of primitive ontology also plays an important role and determines the nature of the link between physics and metaphysics. Any fundamental physical theory always contains a metaphysical hypothesis about what are the fundamental constituents of physical objects. We call this the primitive ontology of the theory (Allori, 2012). The price paid for this approach to developing physical theories is the large number of alternatives that are based on different primitive ontology and nomological variables but with equivalent predictions. Another price paid is the large number of different interpretations of the same theory.

Quantum theory is puzzling in many ways. While the picture of the world that special relativity offers is much more specific in terms of its metaphysical commitments – some of which were discussed in the previous chapter – this is not the case with quantum theory. There is a plethora of interpretations of quantum theory. Below are three popular interpretations from a long list of about 20 of them:

• Copenhagen interpretation

• Causal interpretation

• Many-worlds interpretation

The Copenhagen interpretation of quantum theory is the most popular amongst physicists. This interpretation rests on the principle that the nature of changes in the world is indeterministic and discrete. This interpretation of quantum theory deems all inquiries regarding the state of particles before they are measured to be meaningless. It is the measurement process that selects one possible state from a set of possible states. Therefore, in the context of this interpretation, the reality of what is observed cannot be separated from the act of observation. This interpretation presents us with a radical view that physical reality is a "mind world". It has also given rise to popular speculations that the human mind has the power to change the state of reality. However, nothing of this sort has been proven experimentally.

The causal interpretation is a deterministic theory where faster-than-light interactions are allowed. This interpretation argues that quantum particles have a hidden wave, also called a "pilot wave" that guides their motion. The world is viewed in a holistic sense, i.e., everything interacts with everything else and interactions are virtually instantaneous and are not limited to particles that are in close vicinity but extend to every particle in the universe. This interpretation also presents us with an alternative radical view of the world and is closer to that of Newtonian classical mechanics where gravity emerges due to action-at-a-distance and all bodies in the universe affect each other through it.

The many-worlds interpretation of quantum mechanics asserts that there is an objective reality and that all possible alternatives represent real worlds in an infinite multiverse where everything that could possibly happen has occurred in some of the universes that comprise it. The different universes do not interact with each other and all possibilities exist simultaneously forever. This is an intriguing interpretation of quantum theory that has given rise to popular speculation that, like for example, there could be ways of "jumping" from one universe to another and that all people are wealthy in some parallel universe. However, none of these claims are falsifiable because the parallel universes do not interact and, as a result, they are bold hypotheses, much like the intelligent design hypothesis.

It is important to understand that each different interpretation of quantum theory commits to a different world in terms on structure and operation. At the same time, it remains a puzzle how all these different theories with different interpretations produce the same experimental results. What is the underlying reason for all these different interpretations of the presumably most successful theory of physics? The answer may partly lie in Bell's theorem (Bell, 1987) that implies that every quantum theory must violate either counterfactual definiteness or locality or both but also alludes to the possibility that our freedom of choice to perform one experiment over another is limited and everything is already predetermined. This last possibility is interesting because it brings quantum theory closer to unification with Einstein's relativity, which presents us with a tenseless theory of time and a world in which everything pre-exists, as discussed in the previous chapter.

Counterfactual definiteness is the hypothesis that material objects have existence and possess properties, even if they have not been measured. For example, if one measures the speed of a particle, then one may assume that this measurement is objective. If the speed is measured, then any reference to the position that would have been obtained for a given speed makes sense. Realists about physical theories must adhere to counterfactual definiteness because realism, in its pure form, rests on the assumption that material objects have defined properties whether they are measured or not. Anti-realists do not accept counterfactual definitiveness, in the same way that those that adhere to the Copenhagen interpretation of quantum theory do. Instead, they either adhere to locality or to a pre-existing reality with no freedom of choice.

Locality is the hypothesis that an object is influenced only by its immediate surroundings. Newtonian mechanics did not include this assumption because, in that theory, gravitational forces act instantaneously and at infinite distances. In relativity theory on the other hand, there is an upper limit in the speed of propagation of information that happens to be equal to the speed of light in vacuum and as a result this theory conforms to the hypothesis of locality. The problem that arises is that in quantum theory there is the phenomenon of distant particle entangling, which has been demonstrated experimentally. This phenomenon presents us with a serious problem: action-at-a-distance is not compatible with relativity theory and an upper limit in the speed of information propagation.

Although there are some objections to Bell's theorem (Christian, 2007), the issue of non-locality in quantum mechanics is real. Physicists have been forced to take sides due to this phenomenon. Most reject realism and counterfactual definiteness in favor of locality and relativity. This has also been one cause of the plethora of quantum mechanics interpretations. Some of the interpretations reject counterfactual definiteness and some reject locality. However, both the presence of non-locality and the emergence of properties after they are measured are considered spooky interactions by many. It would be a great relief to them if quantum mechanics were falsified (Harokopos, 2011). On the other hand, there are those who think that relativity will be falsified one day in favor of a probabilistic and non-deterministic world. Thus, the tensions in physics are more intense than they appear to be and progress is slow. The plurality of interpretations is just too confusing to scientists, even if they refuse to admit that. It is also a fact that a large number of physicists are not aware of these foundational conflicts. Science, armed with relativity and quantum mechanics, cannot tell us what kind of world we inhabit. As a matter of fact, physicists are not even sure if these theories are correct since they cannot be fully integrated, although that has been made partly possible in certain domains of operation. Some are even talking about the cause of the failure of unification being the existence of forces outside our universe. Specifically, some have investigated the possibility that gravity is an outside force (Wiithrich, 2004) and that this may be the reason of the failure of unification of general relativity with quantum mechanics.

It is becoming more evident at this point in time in the history of mankind that the picture of a science that slowly progresses towards a theory of everything may no longer be true. For example, our scientific theories cannot tell us how the macroscopic world that is ruled by Newton's second law and with its deterministic, time-reversible, form, discussed in more detail in Chapter Three, autonomously and without any intervention emerges from the probabilistic world of quantum mechanics where spooky interactions take place.

Quantum theory and its plethora of interpretations is an indication that one should view such theory as a mere instrument and not as one with explanatory context. In the case of relativity theory, the situation is a bit different because this theory does not have a plethora of interpretations. This has compelled some to argue that quantum theory is possibly wrong (Lyre, 2010). It is interesting that scientists who have failed to unite the two most important theories of physics, general relativity and quantum physics, make an appeal to the scientific method to oppose intelligent design when they know that for relativity to make sense, the macroscopic world must be a four-dimensional continuum with events, instead of substances, and that for quantum theory to make sense the world must either violate locality or counterfactual definiteness, or even worse, it must be a world devoid of any choice. It is also disturbing that, quite often, one hears the inaccurate claim that quantum mechanics and special relativity theory have been united to a single consistent theory. According to Aleman (2008), this is a widespread assumption but there still remains a considerable amount of unresolved problems to which few pay attention. Or is it maybe that this is another case of scientific obscurantism?

Commitment to properties that emerge only after observation, like in quantum mechanics, is no better than asserting that there is an intelligent cause for everything, including life. The difference up to this point has been that the defenders of emerging scientific dogmas claim falsifiability. The aim of this book is to show how the proponents of an intelligent cause can also claim falsifiability if they commit to a richer interpretation of their principle, which I have called *intelligent interaction* in this book. The objective in this book is not a choice of one dogma over another but to only show that there are alternative ways of looking at our world that should be incorporated into the educational curriculum for the purpose of training future scientists to pursue the further development of these ways and their possible falsification. Only when alternatives clash can science progress. Any effort of obscuring alternatives shows a weakness on the part of those that instrument it.

Section II – INTELLIGENT INTERACTION

"Every truth passes through three stages before it is recognized. In the first, it is ridiculed; in the second, it is opposed; in the third, it is regarded as self-evident" – Arthur Schopenhauer, German Philosopher (1788-1860).

CHAPTER SIX: THE MECHANISM OF INTELLIGENT INTERACTION

The impossibility of motion, and, as a consequence, the impossibility of an autonomous world in a 3-dimensional infinitely divisible space or in a vacuum, was argued since the ancient times by Zeno and his paradoxes of motion, as discussed in Chapter One. In Chapter Two, we saw that Descartes thought that the only possible world with motion and duration is one that is constantly recreated at every instant of time. His followers came up with the doctrine of occasionalism, the hypothesis that all causes are attributed to God. Leibniz attempted to salvage the autonomy of a relational world grounded on the notion of a living force, but failed. Newton came up with his concept of force in an attempt to salvage the reality of this world but he himself admitted that his theory did not suffice for that purpose, as was discussed in Chapter Three. Then, in Chapter Four, special relativity was briefly introduced, the first accurate theory of physics that nevertheless describes a quasi-autonomous but deterministic and eternal four-dimensional world in which the concept of free will, time and change are intriguing but also quite ambiguous. We also discussed some specific inconsistencies in Einstein's relativity when applied to a three-dimensional world. Then, in Chapter Five, we briefly considered the aspects of quantum physics that challenge relativity with spooky interactions and non-local behavior at the particle interaction level. The end result is that after about 450 years of a scientific revolution since the time of Galileo, the same primitive issues that haunted philosophers at the time of Zeno and Plato return to haunt science today. Is this an autonomous world? Can science save the phenomena? Contemporary science cannot provide answers to these important questions because of its foundational problems, some of which were discussed in Chapters One through Five.

From God to mechanism

After the unsuccessful rational approach to the concept of motion by Descartes and the failure of Leibniz to salvage the autonomy of the world, the possibility of an intervening intelligent cause gained popularity. According to Nicolas Malebranche and other seventeenth-century Cartesian occasionalists, what we actually call causes, such as forces that cause acceleration, including the force of gravity, for example, are really no more than occasions on which, in accordance with his own laws, God acts to bring about the effect (Brown, 1995). Thus, according to occasionalists, nature appears to operate based on its own rules and causes but behind the scenes, or in the unobservable background, there is the intelligence of a God that actually causes the phenomena. If one were to replace the notion of God in the occasionalism doctrine by the notion of a mechanism, then a modern – or mechanical – occasionalist could assert that what we actually call causes are no more than occasions on which a mechanism acts to bring about the effect. In this modern version of Cartesian occasionalism, which is compatible with a mechanistic view of the world, time emerges as an ordered progression of instances, or 'nows', on which the interacting mechanism acts to bring about the effect. At the level of the phenomena, time and motion cannot be separated because there is no motion without time and no time without motion, i.e., time and motion are inextricably related, as in relativity theory. Under this model, reality, in the sense of the physical reality we observe, is an effect but also part of the mechanism that creates it. Thus, causality in the observed world is only epiphenomenal, as the assumed mechanism is the real cause of all phenomena.

An epiphenomenon is a secondary phenomenon that occurs in parallel with a primary phenomenon. The primary phenomenon in mechanical occasionalism is the operation of a mechanism and its effect is our physical reality. Generally, the secondary phenomenon cannot affect the operation of the primary phenomenon in the sense of changing its rules of operation. However, in mechanical occasionalism, the secondary phenomenon provides information, in the form of feedback, to the primary phenomenon, and hence to the operations of the mechanism. The rules of the mechanism determine whether the feedback provided by the secondary operation, which is in effect our physical reality, will or must be taken into account. This is the type of functionality that allows some form of free will to emerge but only within the bounds imposed by the operation of the mechanism.

This is the model of a functional virtual reality. It is not a virtual reality or a computer simulation in the traditional sense of these terms. In a functional virtual reality, free interactions are allowed within the limitations of the physical laws imposed by the mechanism. Thus, there is some degree of autonomy and some degree of free will but both are constrained within the limits of physical laws. The future can be determined to a high degree of accuracy if the initial conditions are known, but since free interactions are allowed, local effects are predictable only after their initial conditions emerge.

One could view spacetime in this model as a kind of a substance that acts as a mechanism to create reality, which is the epiphenomenon. This is a different kind of spacetime to that of relativity theory. It is discrete and substantival, in the sense of a digitized supersolid substance of some kind that exists independently of the events that occur in it. Non-locality is a feature of the mechanism to accomplish its task of bringing about the effects in the functional virtual reality. Here is what is happening according to this model: as one moves towards the macrocosmic level, special and general relativity apply because these are the rules that govern the observed reality as dictated by the mechanism. But as one approaches the particle level, quantum physics applies because this is the level at which the mechanism operates at. In order to observe the mechanism, one would have to probe the Planck scale or below that, something that is impossible with our current technology and energy levels available at particle accelerators. However, even if we were able to reach that level, we may be faced with a reality that will be beyond comprehension or measurement. Thus, we may be forced to find alternative indirect means of falsifying or corroborating the hypothesis of functional virtual reality and, as a consequence, the hypothesis of intelligent interaction. That of course would limit the capability of proving the hypothesis true, but this is how physics works. Science without falsification through experimentation is an exercise in futility.

Forces and gravitation as external causes

As a result of the above alternative mechanistic view of the world, all fundamental forces in the universe, including gravitation, have an external cause to the level of the phenomena and spacetime is a substance of some kind that facilitates the energy transfer required for their manifestation and interaction. These ideas may not be completely new as has already been pointed out throughout the book. What is new in this book is an alternative look at the world as opposed to a supposedly autonomous world we have been exposed to since the advent of Newtonian mechanics. Specifically, as will be discussed in the next chapter, the new approach claims that we can ground such metaphysics on the physics of alternative laws of motion based on the notion that power is the cause of motion rather than the Newtonian force (Harokopos, 2005). We further argue that Newtonian mechanics is just a subset of those laws which carried a biased and preferred look at the world from the point of view of an assumed autonomy. The alternative laws based on power fit well with a new foundation of physics according to which all phenomena are caused by energy transfer, including those attributed to gravitation, and with the falsifiable hypothesis that our world is a functional virtual reality manifested by mechanical intelligent interaction.

When one studies the laws of motion that are presented in the next chapter carefully, it will become evident that there were alternatives that were not considered by the scientific establishment when the foundations of modern science were laid, perhaps because of the fact that they offered a richer interpretation of physical reality that challenged the autonomy of this world. The two models that were accepted were first the Newtonian model and then its extension for relativistic speeds and energies, Einstein's relativity. As pointed out in the preceding paragraphs, the Newtonian model was initially presented by the scientific establishment as a solid foundation of an autonomous macro world when not even its creator believed that this was possible, as was detailed in Chapter Three. Einstein's model, special and general relativity, did not challenge the autonomy of the world although it removed causality from it and attributed gravitation to geodesic acausal motion resulting from the geometry of spacetime. Two notorious inconsistencies of special relativity were discussed in Chapter Four along with the implications arising from their resolution. While relativity applies at the macro world, a science for the micro world, quantum physics, was presented as a model of an autonomous world leading to perplexing foundational problems, some of which were briefly mentioned in Chapter Five. The indetermination of quantum theory combined with the measurement problem make it quite possible that this theory is wrong. At no point throughout the history of modern science, was the possibility that this is not an autonomous world taken seriously. I believe that now it is time that this possibility is taken seriously because neither the purported autonomous models nor the acausal models seem to lead anywhere and more importantly, it appears that they cannot be unified.

Below is a summary table of the status of modern science which also illustrates why general relativity and quantum mechanics will never be united without the replacement of one or both theories:

101

Theory	Gravity	Smallest level	Fastest speed
General Relativity (GR)	Present	0	Speed of light c
Quantum Mechanics (QM)	0	Planck level h	∞

Table 6-1: The impossibility of uniting general relativity with quantum mechanics

It may be seen from the above table that in GR the gravitational constant is G (weak – field limit), but in QM this constant is set equal to 0. At the same time, in QM there is the Planck level with a constant equal to h while the value of the constant in GR is 0. Furthermore, the maximum speed of information transmission in GR is equal to the speed of light c but in QM that speed is infinite. As a result, GR and QM cannot be united because any such attempt fundamentally alters the structure of each theory, as both theories have been proven to be accurate independently of the other. It is possible that the obscurantism that prevailed in the case of Newtonian mechanics, as discussed in Chapter Three, has, in this case, been replaced by irrationality. Otherwise, how could anyone explain the scores of scientists who have devoted their lives into uniting GR and QM and why the substantial funds were allocated for a purpose that it is clear just by looking at Table 6-1 that it was doomed?

In the next Chapter, I present the laws of motion based on power and compare them with those of Newton. These laws of motion serve as the foundation of the physics of the alternative view of the world introduced in Chapter Six, based on the principle of intelligent interaction.

CHAPTER SEVEN: POWER AS THE CAUSE OF MOTION

The failure of modern theories to provide solutions to the foundational problems of classical mechanics is partly due to the lack of rigid foundations. Quantum uncertainty and the four-dimensional spacetime of relativity, discussed in Chapters Five and Chapter Four, respectively, have now taken the place of the determinism and of the unobservable absolute space and universal time of classical mechanics. Mysterious action-at-a-distance prevails in the quantum world, and attempts to quantize gravity and unite quantum mechanics with general relativity have failed to this date and will probably never succeed as was argued at the end of the previous chapter. The alternative system of laws of motion based on power presented in this chapter is part of an effort to offer a new foundation of physics, starting at the classical level, which can provide an alternative approach for answers to unsolved problems of modern science through new and different metaphysical commitments about the nature of reality.

It is quite intriguing that unlike their seventeenth-century counterparts, most contemporary physicists and philosophers of science think of classical mechanics as unproblematic. Butterfield (2004) mentions two errors found in this view that correspond to what he calls 'the matter-in-motion picture' and 'the particle-in-motion picture'. According to the matter-in-motion picture, for example, bodies are collections of particles separated by voids that can move in vacuum and interact with each other, while their motion is completely determined by Newton's second law. This view has become a part of an educated layperson's common sense nowadays, but according to Butterfield it is problematic; amongst other things, it does not offer any explanation of the mechanism(s) of the assumed interactions but resorts to concepts such as forces acting across an intervening void ('action-at-a-distance').

In a similar way to the matter-in-motion picture, the notion of force has also become part of an educated layperson's common sense, thanks to the empirical support the laws of mechanics have enjoyed over the past 300 years. It is well-known, however, that Newton was heavily criticized for his use of the notion of force in an effort to ground his physics on his metaphysics and there is still considerable interest in the metaphysics of his Principia. In Science and Hypothesis, Poincaré (1952) writes:

> When are two forces equal? We are told that it is when they give the same acceleration to the same mass, or when acting in opposite directions they are in equilibrium. This definition is a sham.

In Principles of Dynamics, Donald T. Greenwood (1965) offers an introduction to the issues raised by Newton's concept of force:

> The concept of force as a fundamental quantity in the study of mechanics has been criticized by various scientists and philosophers of science from shortly after Newton's enunciation of the laws of motion until the present time. Briefly, the idea of a force, and a field force in particular, was considered to be an intellectual construction, which has no real existence. It is merely another name for the product of mass and acceleration, which occurs in the mathematics of solving a problem. Furthermore, the idea of force as a cause of motion should be discarded since the assumed cause and effect relationship cannot be proved. (Emphasis added)

I will now proceed by presenting the axiom and laws of motion based on power (Harokopos, 2005). The details of the derivation can be found in the referenced paper.

The axiom of motion

The axiom of motion is an expression that relates the velocity and the time rate of change of momentum of a particle to a scalar quantity called the time rate of change of kinetic energy, also known as (instantaneous) power. The status of this axiom is assumed here to be that of an a priori truth as opposed to a self-evident or empirical principle.

Axiom of motion: *The time rate of change of the kinetic energy of a particle (power) is equal to the scalar product of its velocity and time rate of change of its momentum.*

The laws of motion

Law of inertia: *If the time rate of change of kinetic energy (power) remains zero, a particle maintains its state of rest or its state of uniform motion in a straight line or its state of curvilinear motion at constant speed.*

Law of interaction: *To every action there is an equal and opposite reaction; that is, in an isolated system of two particles acting upon each other, the mutual time rate of change of kinetic energies (mutual powers) are equal in magnitude and opposite in sign.*

Law of universal gravitation: *All particles under the influence of gravity move in such a way as for the time rate of change of their kinetic energy (power) to be equal to (minus) the time rate of change of their potential energy.*

Comments on the axiom and laws of motion

The laws of motion presented above can be considered as an alternative to Newton's laws of motion. Thus, one should refrain from evaluating these laws in the context of Newtonian mechanics, since the two systems of laws make commitments to different metaphysics, especially, in a different notion of causation.

The alternative to Newton's laws given by the laws of motion based on power allows resolving a few notorious philosophical problems in the foundations of classical mechanics and, particularly, the need to consider fictitious causes when the second law is used in non-inertial reference frames. In the case of observers at rest in accelerated reference frames in either rectilinear or uniform circular motion, the time rate of change of kinetic energy is zero in that moving frame and thus no additional fictitious power cause is needed to explain the state of motion. Again, this is only true if power is defined as the cause of motion. If force is defined as the cause of motion, then in both of the non-inertial reference frames mentioned, fictitious causes must be present. Specifically, in the case of rectilinear motion, observers at rest in an accelerated frame must assume inertial forces acting, and in the case of observers at rest in a uniformly rotating reference frame, it must be assumed that centrifugal forces are acting.

The same conclusion holds in the case of fictitious Coriolis forces acting on freely moving particles in rotating reference frames. In this case, since such fictitious forces are always orthogonal to the velocity of a particle in motion, then, for rotating observers with a particle, it turns out that the time rate of change of kinetic energy of the particle is equal to zero. The same result is true for observers at rest since, in that case, the time rate of change of momentum of a freely moving particle is zero. Note that fictitious forces need to be considered regardless of whether force or power is defined as the cause of motion when a dynamic analysis is carried out. However, when power is defined as the cause of motion, there are no philosophical issues arising from the need to consider fictitious causes of motion in non-inertial reference frames and this is the point that was just made. Again, this is due to the metaphysical commitment of the alternative laws of motion presented in this book that the cause of motion is power and not force. Thus, the transition from force to power as the cause of motion leads to a compatibility with the epistemological principle, which states that every phenomenon is to receive the same interpretation from any given moving coordinate system. This epistemological principle also plays an important role in the axiomatic foundation of the theory of relativity (Reichenbach, 1958).

Axiom of motion: *The time rate of change of the kinetic energy of a particle (power) is equal to the scalar product of its velocity and time rate of change of its momentum.*

Let us recall that the mathematical expression of Newton's second law, after a suitable choice of units is made, is as follows:

$$\mathbf{F} = \frac{d\mathbf{p}}{dt} = \frac{d}{dt}(m\,v) \quad (7-1)$$

With his second law, Newton defined the force **F** that affects a change in momentum as the cause of motion and equated it to the time rate of change of momentum **p**. The difficulties in deriving this form of Newton's second law from his original work in the Principia were discussed in Chapter Three. The laws of motion presented in this book, based on the axiom of motion, challenge the notion that the Newtonian force is the metaphysical cause of motion and a true foundation of mechanics. In these laws of motion, the metaphysics of force are replaced by those of the time rate of change of kinetic energy, also known as power. In a way analogous to Newton's second law, the axiom of motion can be expressed as follows:

$$P = \frac{d(E_k)}{dt} \quad (7-2)$$

where P is the (instantaneous) power and E_k the kinetic energy of a particle. Notice that this is a scalar law of motion as opposed to Newton's second law which is a vector law. In this law of motion, the power P is a scalar quantity as opposed to force **F** which is a vector quantity.

Law of inertia: *If the time rate of change of kinetic energy (power) remains zero, a particle maintains its state of rest or its state of uniform motion in a straight line or its state of curvilinear motion at constant speed.*

In the case of Newtonian mechanics, a centripetal force must apply for a particle to move at constant speed and remain in circular path and its effect is a centripetal acceleration. On the contrary, in the context of the laws of motion based on power presented in this chapter, when a particle is on a uniform circular path no cause is required to maintain its state of motion because power, the assumed cause of motion, is zero. The particle maintains its state of motion due to inertia alone, although a force applies and it is always orthogonal to velocity.

Inertial path possibilities compatible with the laws of motion based on power include motion in 8-shaped figures and cycloid paths, just to mention a few possibilities. However, some of these paths may represent physical possibilities and others may not. Uniform circular motion is a physical possibility in both micro and macro scales and this has been confirmed empirically. The choice of specific curvilinear motions over others as an effect of inertia, if power is postulated to be the cause of motion, is made by the mechanism discussed in Chapter Six through intelligent interaction. The law of inertia presented here is a statement that the state of such motions is maintained in the absence of a cause, if power is assumed to be the cause of motion. However, the law does not provide a justification for the existence or preference of certain states of motions over others in the absence of a cause of motion.

It is known from differential geometry that if a curve is regular, then re-parametrization exists such that the curve has unit speed (O'Neill, 1997). Thus, any particle can be made to move with constant speed along any curve in space using proper arc-length re-parametrization resulting in constant kinetic energy and as a consequence, zero power.

The law of inertia is a statement about the tendency of particles to maintain their state of motion at rest, or of uniform motion, or of curvilinear motion at constant speed, when power remains zero, and this tendency is called inertia. Again, this law of inertia is based on the metaphysical hypothesis that power is the cause of motion. A consequence from such hypothesis is that the set of cause-free paths now includes all paths where the kinetic energy remains constant, instead of just uniform rectilinear motion and the state of rest defined in Newtonian mechanics.

Thus, I essentially argue that Newton's First Law only considers just two possibilities from a broader range of possibilities mandated by a more general principle of inertia presented in this chapter. Actually, Poincaré has already suspected that such a more general law exists and he wrote (Poincaré, 1952):

> The Principle of Inertia. – A body under the action of no force can only move uniformly in a straight line. Is this a truth imposed on the mind à priori? If this be so, how is it that the Greeks have ignored it? How could they have believed that motion ceases with the cause of motion? Or, again, that every body, if there is nothing to prevent it, will move in a circle, the noblest of all forms of motion?

109

> If it be said that the velocity of a body cannot change, or there is no reason for it to change, may we not just as legitimately maintain that the position of a body cannot change, or that the curvature of its path cannot change, without the agency of an external cause? Is, then, the principle of inertia, which is not an à priori truth, an experimental fact? Have there ever been experiments on bodies acted on by no forces? And, if so, how did we know that no forces were acting?

Poincaré continues with his discussion of the principle of inertia by stating that Newton's First Law could be the consequence of "a more general principle, of which the principle of inertia is only a particular case."

Law of interaction: *To every action there is an equal and opposite reaction; that is, in an isolated system of two particles acting upon each other, the mutual time rate of change of kinetic energies (mutual powers) are equal in magnitude and opposite in sign.*

In the scholium following the laws of motion, Newton attempts to provide additional support for the third law through a host of observations related to various modes of mechanical interaction between bodies. Some interesting conclusions can be drawn from the closing comments in the scholium (Newton, 1952):

> …But to treat of mechanics is not my present business. I was aiming to show by those examples the greater extent and certainty of the third Law of Motion. *For if we estimate the action of the agent from the product of its force and velocity* and likewise the reaction of the impediment from the product of the velocities of its several parts, and the forces of resistance arising from friction, cohesion, weight, and acceleration of those parts, the action and reaction in the use of all sorts of machines will be found always equal to one another. And so far as the action is

propagated by the intervening instruments, and at last impressed upon the resisting body, the ultimate action will be always contrary to the reaction. (Italics added)

It is clear that Newton was well aware of the product of velocity and force being a measure of action and of reaction, as defined in the law of interaction above. Actually, Newton made use of the law of interaction in his scholium above to justify some particular situations where his third law of action-reaction does not apply directly. But why is it the case that Newton stated his third law in terms of forces and not in terms of the product of force and velocity that he mentions in his scholium quoted above? Why does it appear that a more general law was used to justify some particular situations Newton's third law does not directly apply to, but the latter was stated as a law of mechanics? The answer may be found in the attempt to model gravitation in Newtonian mechanics as the effect of mutual attraction caused by central forces acting at a distance. The third law had to be stated in terms of the mutual action-reaction forces being equal in magnitude and opposite in direction to comply with the particular form of Newton's law of universal gravitation. But again, the third law fails the requirement set forth by the rules of Newton's (experimental) philosophy, for it being deduced from the phenomena; it is just another axiom that must be accepted without proof. Forces acting on different bodies, and especially celestial ones, cannot be experimentally determined to be equal. Only forces acting on the same body can be determined to be equal by experiment.

Law of universal gravitation: *All particles under the influence of gravity move in such a way as for the time rate of change of their kinetic energy (power) to be equal to (minus) the time rate of change of their potential energy.*

With this law of gravitation I argue that Newton's law of universal gravitation is in fact a statement about the form of the potential function $E_p(r)$, which for two point masses, M and m, has the following form:

$$E_p(r) = -\frac{GMm}{r} \quad (7-3)$$

where G is the gravitational constant and r the magnitude of the vector distance r between M and m. Equation 7-3 is an expression of the gravitational potential energy of a mass m in the gravitational field of a mass M.

If we postulate then that energy transfer affects all particles in gravitational motion, but no actual energy transfer occurs between the particles, this can lead to the hypothesis that gravitation is the result of an interaction between all bodies in motion with some (unobservable) substance, in accordance with above equation. Substantival spacetime can serve the role of this substance and can facilitate the interactions in such a way that all spatiotemporal quantities evolve according to certain rules giving rise to the well-known potential function $E_p(r)$ discovered by Newton. This is the essence of the intelligent interaction as discussed in Chapter Six.

I have thus presented an alternative system of laws of motion in the chapter based on the notion that power is the cause of all phenomena, including gravitation. These laws represent a significant departure from the notion that forces are the cause of phenomena, which was adopted by the educational establishment since Newton published his laws of motion. I have already argued in this book that the direction that science took after the publication of Newton's laws heavily depended on the metaphysical commitments made by them. The laws based on power free us from the notion that forces are the causes of the phenomena and allow us to postulate an alternative nature of physical reality based on the notion of intelligent interaction via energy transfer. Along with this new view of reality comes a perspective of the world that can potentially lead to a development of a system of values that can serve as a balance to the system of values cultivated so far by the embraced metaphysics. This is the ultimate aim of this book and there is nothing to hide about it. I am not proposing a direct replacement of metaphysics but a balance of metaphysics. This is why I argue that the notion of intelligent interaction and the possibility that our world is not autonomous but that it is manifested by the interactions of an intelligent mechanism and the constituents of the phenomena should be included in educational curriculums as a viable alternative until it is conclusively falsified. In the next chapter, I will present a starting ground for such falsification.

CHAPTER EIGHT: ALTERNATIVE FOUNDATIONS

Let us recall the main objective of this book as stated in the introduction:

> Therefore, I argue that it may be possible to convince the scientific community to investigate alternative theories that incorporate a suitable variation of the intelligent design hypothesis as a postulate. This could be achieved after demonstrating, or actually reminding people I should say, of the foundational problems of science. Then, go one step further and show how a new theory can offer an alternative view of our world and make new predictions that are testable and falsifiable. This is the strategy followed in this book. But in the process of achieving this goal, it will be necessary to consider intelligent design as part of a postulate that is much broader in scope, called an intelligent interaction. In a nutshell, this new postulate asserts that the universe, the laws of physics and life exist because there is intelligent interaction between a universal mechanism and the constituents of the phenomena. This mechanism is part of the physical universe but it is not directly observable, in the same way that spacetime is not directly observable.

In the first section of this book, in Chapters One through Five, it was shown that there is high ambiguity in the concepts of modern science mainly due to unsolved foundational problems. Although modern science has developed several useful models that allow making accurate predictions about the phenomena, it has alienated people in their search for answers regarding the nature of reality. Foundations have become much more complex and ambiguous.

In Chapter Six, I described the concept of intelligent interaction and the alternative model of physical reality that it offers. Then in Chapter Seven, I presented an alternative system of laws of motion based on the notion that power is the cause of all phenomena, including gravitation. I argued that the physics of these alternative laws of motion is compatible with a functional virtual reality generated and ruled by intelligent interaction.

Arguments leading to provocative hypotheses, like the functional virtual reality hypothesis in this book, for example, are usually treated by the educational establishment at the level of epistemological skepticism. However, as McCabe (2004) has argued, the hypothesis that our universe is part of a computer simulation implementation generates empirical predictions and it is therefore a falsifiable hypothesis. The hypothesis of the existence of external causes to the world of observable phenomena must not be rejected a priori on the basis that it leads to the provocative virtual reality hypothesis but, instead, experimental physics must seriously pursue its falsification or corroboration. Although such a task is highly challenging, the state-of-the-art in precision instrumentation has reached such levels as to allow the initiation of a program of this nature.

Experimental corroboration of the alternative

The objective of an experiment in the context of intelligent interaction will be to falsify or corroborate the functional virtual reality model. The basic premise for conducting such an experiment rests on the hypothesis that a mechanism which continuously recreates a functional virtual reality has an upper limit in its processing capability at the local level (Harokopos, 2011). At large scales, the limit may be practically infinite due to the distributed nature of the mechanism. Thus, given the premise, the objective of the experiment will be to observe local phenomena where there is a deviation from expected results due to a demand of a processing capability of the mechanism that is higher than possible. One possible way of achieving this test is by detecting deviations in the output of a nanoscale processor accelerated at high speeds (ibid.). The processor may consist of a dense array of billions of nanoscale transistors and it will execute a complex set of instructions. Recent developments in the area of single atom transistors can contribute towards this goal. The processor could be made three-dimensional for greater density by utilizing three-dimensional wafer technology. It should be shielded to avoid interference. When this devise is accelerated at high speeds for short distances, deviations may occur from the expected output if there is a limitation in the capability of the hypothesized mechanism to bring about effects. Of course, the processor should be first tested at the same speeds without running any set of instructions in order to check for malfunctioning and presence of noise.

The above was a sketch of a possible experiment for testing the processing capability of a mechanism that continuously recreates a digital functional virtual reality. Other similar experimental setups are possible. For example, devices can be developed that will execute a set of instructions during the re-entry of space vehicles to the Earth's atmosphere from space. During such re-entry, high speeds are achieved allowing the capability of a hypothesized mechanism to maintain causation to be tested on a small and local scale. Again, we are interested in falsification or corroboration of the functional virtual reality hypothesis. Proving this hypothesis true is impossible within this class of experiments.

What is at stake here?

There is something more important than physics or metaphysics that warrants serious consideration of an alternative look at our world: it is the system of values cultivated in people by a science that has placed too much emphasis on a purported autonomy of the material world. The results of the cultivation of this system of values are evident in our rapidly growing technological society, which is dominated by greed and lack of respect for the environment and human values. Even if we are not certain that the alternative look of the world presented in this book is correct, we should still introduce it, or some suitable version of it, into the educational curriculum for the purpose of providing a balance of values. Some would argue that this should be the task of ethics or religion courses without a need for providing alternative theories about the nature of physical reality. However, as I have already pointed out in this book, these are the not subjects that shape value systems any longer in a technologically and material society.

The lack of alternatives to current paradigm makes the risks too high in the sense that in the remote possibility that it is ever proven that our world is not autonomous but there is another level of reality which creates it, there will be a major shock to human perceptions and values. In other words, people should be prepared for the possibility that one day we may somehow discover that this world is the result of some sort of intelligent interaction and that life was not created by random chemical reactions but that some form of design was or is continuously involved at another level of reality. I personally tend towards to being agnostic on this issue, although as I have already pointed out, I am sympathetic towards the hypothesis of intelligent interaction but not a fanatic proponent of it. My objective in this book was to present an alternative to the established paradigm that can satisfy the requirements of the scientific method. This was a preliminary and small step towards this direction. I think the scientific and educational establishment should be glad that there is some work to be done along these lines and I am sure that if the theory has no merit it will be soon falsified and I will be the first to concede to the autonomy of the world. A main message of this book was that intelligent design, the doctrine that human life was designed, can fit into a larger

scheme of an alternative theory about the nature of our reality that it is falsifiable.

Efthimios Harokopos

REFERENCES

Aleman, R. (2008):, "Quantum Mechanics versus Special Relativity: A forgotten conflict", e-print, Pittsburgh Archives, No. 00004313.

Allori, V. (2012): "On the Metaphysics of Quantum Mechanics". Preprint, URL: http://philsci-archive.pitt.edu/9343/

Baker, D. (2004): "Spacetime Substantivalism and Einstein's Cosmological Constant." Proc. Phil. Sci. Assoc., 19th Biennial Meeting, PSA2004.

Barbour, J. B. (1999): The End of Time, New York: Oxford University Press.

Barbour, J. B. (2001): The Discovery of Dynamics: A Study from a Machian Point of View of the Discovery and the Structure of Dynamical Theories. New York: Oxford Univ. Press.

Barnes, J. (1979): The Presocratic Philosophers, London: Routledge.

Behe, M. J. (1996): Darwin's Black Box: The Biochemical Challenge to Evolution, New York: Free Press.

Bell, J. (1987): Speakable and Unspeakable in Quantum Mechanics, Cambridge: Cambridge University Press.

Benacerraf, P. (1962): "Tasks, super-tasks, and the modern Eleatics", The Journal of Philosophy 59, pp. 765-784.

Brown, S. (1995): "The Seventeenth-Century Intellectual Background", N. Jolley, ed. in The Cambridge Companion to Leibniz, Cambridge: Cambridge University Press. Pp. 43-66.

Butterfield J. (2004): "Between Laws and Models: Some Philosophical Morals of Lagrangian Mechanics", 2004. e-print, Pittsburgh Archives, No. 00001937.

Christian, J. (2007): "Disproof of Bell's Theorem: Further Consolidations", arXiv:0707.1333

Dembski, W. A. (2002): No Free Lunch: Why Specified Complexity Cannot Be Purchased Without Intelligence, Lanham: Rowman & Littlefield.

Einstein, A. (1905): "Zur Elektrodynamik bewegter Korper", Annalen der Physik", vol. 322, pp. 891–921.

Einstein, A. (1916): Relativity: The Special and General Theory, New York: H. Holt and Company.

Esfeld, M. (2009): "Hypothetical metaphysics of nature", in Michael Heidelberger and Gregor Schiemann (eds.), The significance of the hypothetical in the natural sciences, Berlin: de Gruyter. Pp. 341–364.

Franco, A. (2001): "Duration and Motion in a (Cartesian) World which Is Created Anew at Each Moment by an Immutable and Free God" Critica, vol. 33, No. 99, pp. 19–45

Garber, S. D. (1995): "Leibniz: physics and philosophy", N. Jolley, ed. in The Cambridge Companion to Leibniz, Cambridge: Cambridge University Press. Pp. 270-352.

Greenwood, D. T. (1965): Principles of Dynamics, New Jersey: Prentice-Hall. Pp. 23-25.

Guthrie, W. K. C. (1965): The History of Greek Philosophy, Cambridge: Cambridge University Pres.

Harokopos, E. (2005): "Power as the Cause of Motion and a New Foundation of Classical Mechanics", Progress in Physics, vol. 2, pp. 82-91.

Harokopos, E. (2011): "A Functional Virtual Reality", FQXi Essay Contest, URL: http://www.fqxi.org/community/forum/topic/846

Hey T. & Walters P. (2009): The New Quantum Universe, Cambridge University Press.

Kirk, G. S., Raven, J. E. and Schofield, M. (1983): The Presocratic Philosophers, Cambridge: Cambridge University Press.

Laudan, L. (1981): "A Confutation of Convergent Realism", Philosophy of Science, vol. 48, No. 1, pp. 19-49

Lyre, H. (2010): "Why Quantum Theory is Possibly Wrong", Foundations of Physics, 40(9), pp. 1429-1438.

McCabe G. (2004): "Universe Creation on a Computer", e-print, Pittsburgh Archives, No. 00001891.

McLaughlin, W. and Miller, S. L. (1992): "An epistemological Use of Nonstandard Analysis to Answer Zeno's Objections Against Motion", Synthese 92, pp. 371-384.

Meirovitch S. L.(1970): Methods of Analytical Dynamics, New York: McGraw – Hill.

Milgrom, M. (2002): "Do Modified Newtonian Dynamics Follow from the Cold Dark Matter Paradigm?", Astrophysical Journal, May 2002.

Mostratou, S. (1948): The Other Heroes: The First Greek Philosophers (in Greek: Μοστράτου Σμαράγδα , Οι Άλλοι

Ήρωες: Οι Πρώτοι Έλληνες Φιλόσοφοι), Athens: Aetos S.A. (All translations in the text are mine)

Newton, I. (1952): Mathematical principles of natural philosophy. Trans. by Andrew Motte and rev. by Florian Cajori, R. M. Hutchins, ed. in Great Books of the Western World: 34. Newton Huygens. Encyclopaedia Britannica, Chicago.

O'Neill B. (1997): Elementary Differential Geometry, San Diego: Academic Press.

Poincaré, H. (1952): Science and Hypothesis, New York: Dover.

Reichenbach, H. (1958): The Philosophy of Space and Time, New York: Dover.

Roberts J. T. (2003): "Leibniz on force and absolute motion". J. Phil. Sci., v. 70, pp. 553-571.

Shamsi, F. A. (1994): "A Note on Aristotle, Physics 239b5-7: What Exactly Was Zeno's Argument of the Arrow?", Ancient Philosophy 14, pp.51-72.

Sklar, L. (1997): Space, Time and Space-Time, Berkeley: University of California Press.

Svozil, K. (1995): "Set Theory and Physics", Foundations of Physics, 25, pp. 1541-1560.

Vickers P. (2009): "Was Newtonian Cosmology Really Inconsistent?", in Studies In History and Philosophy of Modern Physics, vol. 40, Issue 3, Elsevier, Pp. 197-208.

Weinberg, S. (1989): "The Cosmological Constant Problem". Reviews of Modern Physics 61, pp. 1-23.

Weatherall, J. O. (2011): "On (Some) Explanations in Physics", Philosophy of Science, vol. 78, No. 3, pp. 421-447.

Wiithrich, C. (2004): "To Quantize or Not to Quantize: Fact and Folklore in Quantum Gravity", Proc. Phil. of Sci. Assoc., 19th Biennial Meeting, PSA2004 2004.

INDEX

I

K

L

M

N